T0275694

Scaling Chemical Processes

Scaling Chemical Processes

Practical Guides in Chemical Engineering

Jonathan Worstell
University of Houston
Worstell and Worstell, Consultants

AMSTERDAM • BOSTON • HEIDELBERG • LONDON
NEW YORK • OXFORD • PARIS • SAN DIEGO
SAN FRANCISCO • SINGAPORE • SYDNEY • TOKYO
Butterworth-Heinemann is an imprint of Elsevier

Butterworth-Heinemann is an imprint of Elsevier
The Boulevard, Langford Lane, Kidlington, Oxford OX5 1GB, UK
50 Hampshire Street, 5th Floor, Cambridge, MA 02139, USA

Copyright © 2016 Elsevier Inc. All rights reserved.

No part of this publication may be reproduced or transmitted in any form or by any means, electronic or mechanical, including photocopying, recording, or any information storage and retrieval system, without permission in writing from the publisher. Details on how to seek permission, further information about the Publisher's permissions policies and our arrangements with organizations such as the Copyright Clearance Center and the Copyright Licensing Agency, can be found at our website: www.elsevier.com/permissions.

This book and the individual contributions contained in it are protected under copyright by the Publisher (other than as may be noted herein).

Notices
Knowledge and best practice in this field are constantly changing. As new research and experience broaden our understanding, changes in research methods, professional practices, or medical treatment may become necessary.

Practitioners and researchers must always rely on their own experience and knowledge in evaluating and using any information, methods, compounds, or experiments described herein. In using such information or methods they should be mindful of their own safety and the safety of others, including parties for whom they have a professional responsibility.

To the fullest extent of the law, neither the Publisher nor the authors, contributors, or editors, assume any liability for any injury and/or damage to persons or property as a matter of products liability, negligence or otherwise, or from any use or operation of any methods, products, instructions, or ideas contained in the material herein.

Library of Congress Cataloging-in-Publication Data
A catalog record for this book is available from the Library of Congress.

British Library Cataloguing-in-Publication Data
A catalogue record for this book is available from the British Library.

ISBN: 978-0-12-804635-7

For Information on all Butterworth-Heinemann publications
visit our website at http://www.elsevier.com/

Working together to grow libraries in developing countries

www.elsevier.com • www.bookaid.org

Publisher: Joe Hayton
Acquisition Editor: Fiona Geraghty
Editorial Project Manager: Maria Convey
Production Project Manager: Anusha Sambamoorthy
Cover Designer: MPS

Typeset by MPS Limited, Chennai, India

DEDICATION

To our grandsons

Jaden H. Worstell

Jacob H. Worstell

Jack H. Worstell

CONTENTS

CHAPTER 1

Introduction

INTRODUCTION

Scaling, up and down, is the *raison de etre* for chemical engineering. Scaling chemical processes provided the primary impetus for the development of chemical engineering, which became a separate professional discipline during the first two decades of the 20th century. Prior to those decades, chemists scaled their laboratory experiments into commercial-sized equipment themselves. By scaling, we mean they used larger pots in which to conduct their reactions. With each successively larger pot, they obtained a larger quantity of product. They continued this exercise until the process became heat transfer limited. If product demand increased beyond the capacity of the original commercial-sized pot, the chemist simply built a second, duplicate pot. This replication process continued so long as product demand increased.

These scaling efforts were generally successful since they were batch processes. However, international trade increased dramatically with the development of steel-clad, steam-driven ships. This increase in international trade expanded numerous markets, one being the market for chemicals. To supply the expanding chemicals market, chemists needed more pots, larger pots, or shorter reaction times. Installing more pots would increase capacity, but doing so would not increase pot efficiency or productivity. Installing larger pots would increase pot efficiency and productivity, but doing so would involve transferring significant amounts of energy, as heat, which was the provenance of mechanical engineers. Increasing the pressure and/or temperature at which they conducted a particular reaction would achieve their goal of more product, faster, but operating at elevated pressure and temperature required a knowledge of material strengths and durability, which, again, was the provenance of mechanical engineers.

The advent of the automobile occurred at the same time when international trade increased. The development of the automobile with its

Scaling Chemical Processes.
© 2016 Elsevier Inc. All rights reserved.

internal combustion engine increased the demand for refined petroleum products. During the late 19th century, kerosene was the main product of petroleum refineries. Kerosene is a "middle distillate," that is, it separates from petroleum at a higher boiling point than gasoline and at a lower boiling point than gas oil and lubricating oil. During the 19th century, chemists produced kerosene via batch distillation. However, by the end of the first decade of 1900s, gasoline, a lower boiling component of crude petroleum, had become the primary product of petroleum refiners. The ever-increasing demand for gasoline led to the construction and commissioning of the first continuous refinery in 1912, which required pumps, heat exchangers, and flow, pressure, and temperature control mechanisms, all areas pertinent to mechanical engineering. Therefore, during the last decade of the 19th century and the first decade of the 20th century, chemists began working closely with mechanical engineers to increase the efficiency and productivity of chemical processes and to design continuous operating petroleum refineries.

While the interaction between chemists and mechanical engineers was productive, it was not efficient: chemists had to teach mechanical engineers the chemistry involved in each process or product being developed and mechanical engineers had to teach chemists the pertinent engineering. Both these activities required time, which slowed the development of particular processes and products.

It was the development and commercialization of the Haber ammonia process, which involves chemical equilibrium, catalysis, and high pressure and temperature that clearly demonstrated the need for chemist-mechanical engineers or for mechanical engineer-chemists. Several universities in a number of countries responded to this need by creating programs in "chemical engineering." The newly developed discipline of physical chemistry formed the foundation for chemical engineering.

Thus chemical engineering as a separate discipline arose from the need to scale chemical processes into ever larger pots.

PROCESS SCALING

Chemical engineers do not just scale into larger equipment. They also scale into smaller equipment. Downscaling occurs when an existing commercial plant has significant operating problems that are not amenable to calculated resolution. In other words, experiments must be performed to fully understand and solve the problems. Such testing cannot generally be done in a commercial-sized plant. In this case the chemical engineer will downsize the problem into smaller equipment where the necessary testing can be conducted.

So, how can a chemical engineer be sure that he or she has upscaled a process or downscaled a problem accurately?

To answer that question, let us consider the flow of an incompressible, viscous fluid in a pilot plant and in a commercial plant. We assume the simplest geometry for our example, namely, rectangular coordinates with flow in the x direction only. We designate the pilot plant as the "model" and the commercial plant as the "prototype."

The Navier–Stokes equation for our model is

$$\frac{\partial v_{\mathrm{M}}}{\partial t} + v_{\mathrm{M}}\frac{\partial v_{\mathrm{M}}}{\partial x} = -\frac{1}{\rho}\frac{\partial p_{\mathrm{M}}}{\partial x} + g + \frac{\mu}{\rho}\left(\frac{\partial^2 v_{\mathrm{M}}}{\partial x_{\mathrm{M}}^2}\right) \tag{1.1}$$

where v_{M} is fluid velocity in the x direction (m/s); t is time (s); x is the distance or location coordinate (m); ρ is fluid density (kg/m^3); p is pressure (kg/m $\cdot$ s^2); and μ is fluid viscosity (kg/m $\cdot$ s). The Navier–Stokes at an equivalent location in our prototype as in the model is

$$\frac{\partial v_{\mathrm{P}}}{\partial t} + v_{\mathrm{P}}\frac{\partial v_{\mathrm{P}}}{\partial x} = -\frac{1}{\rho}\frac{\partial p_{\mathrm{P}}}{\partial x} + g + \frac{\mu}{\rho}\left(\frac{\partial^2 v_{\mathrm{P}}}{\partial x_{\mathrm{P}}^2}\right) \tag{1.2}$$

Note that all prototype variables have the same units as their equivalent model variables.

For flowing fluid in the prototype to behave similarly to the flowing fluid in the model, one equation, independent of size, must be valid for both processes. Thus we need a common set of variables for the prototype process and the model process; that is, we need to derive a size independent Navier–Stokes equation.

To derive the size independent Navier–Stokes equation, note that the model and the prototype differ in size by the ratio

$$\frac{x_{\mathrm{P}}}{x_{\mathrm{M}}} \tag{1.3}$$

In Model Theory this ratio is always greater than one; thus it is always the prototype variable divided by the model variable. The above ratio (1.3) is dimensionless. We designate it as

$$\frac{x_P}{x_M} = \Pi_x \tag{1.4}$$

where the subscript x on Π denotes a ratio of coordinate or location variables. We make this designation because we can also determine characteristic lengths for the prototype and the model that equal Π_x, namely

$$\frac{L_P}{L_M} = \Pi_x \tag{1.5}$$

where L_P and L_M are the characteristic lengths for the prototype and model, respectively.

Combining Eqs. (1.4) and (1.5) gives us

$$\frac{x_P}{x_M} = \frac{L_P}{L_M} = \Pi_x \tag{1.6}$$

Multiplying by x_M and dividing by L_P yields

$$\frac{x_P}{L_P} = \frac{x_M}{L_M} = \left(\frac{x_M}{L_P}\right)\Pi_x \tag{1.7}$$

which provides us with a dimensionless coordinate x^*, namely

$$x^* = \left(\frac{x_M}{L_P}\right)\Pi_x \tag{1.8}$$

Thus

$$\frac{x_P}{L_P} = \frac{x_M}{L_M} = x^* \tag{1.9}$$

Separating the equalities gives us

$$\frac{x_M}{L_M} = x^* \quad \text{and} \quad \frac{x_P}{L_P} = x^* \tag{1.10}$$

Thus we have reduced two location variables to one dimensionless location variable. Reducing all the variables in the two Navier–Stokes equations to dimensionless variables permits us to write one Navier–Stokes equation that describes the flow behavior in both the prototype and the model.

Applying the same reasoning to fluid velocity yields

$$\frac{v_M}{V_M} = v^* \quad \text{and} \quad \frac{v_P}{V_P} = v^* \tag{1.11}$$

where v_M and v_P are the fluid velocities in the model and the prototype, respectively, and V_M and V_P are the characteristic velocities for the model and the prototype, respectively. v^* is dimensionless velocity, which is equal to $(v_M/V_P)\Pi_v$.

We can derive a dimensionless time from the definition of velocity, which is

$$v = \frac{x}{t} \tag{1.12}$$

Rearranging this definition gives us

$$t = \frac{x}{v} \tag{1.13}$$

We can establish dimensionless times for the model and prototype by noting that

$$t_M = \frac{x_M}{v_M} = \frac{L_M x^*}{V_M v^*} \tag{1.14}$$

but

$$t^* = \frac{x^*}{v^*} \tag{1.15}$$

Substituting $t^*$for x^*/v^*, we get

$$t_M = \left(\frac{L_M}{V_M}\right) t^* \tag{1.16}$$

We can use the same reasoning for the prototype to arrive at its dimensionless time, which is

$$t_P = \left(\frac{L_P}{V_P}\right) t^* \tag{1.17}$$

We define the pressure of the model as

$$p_M = \frac{F_M}{A_M} \tag{1.18}$$

where F_M designates the force impinging upon an area ($kg \cdot m/s^2$) and A_M as the area experiencing the force (m^2). Thus the units of pressure are $kg/m \cdot s^2$. Using Newton's second law to determine F_M, we obtain

$$p_M = \frac{F_M}{A_M} = \frac{m(dv_M/dt_M)}{x_M^2} \tag{1.19}$$

Converting v_M, t_M, and x_M to their dimensionless equivalents gives us

$$p_M = \frac{m(d(V_M v^*)/d(L_M/V_M)t^*)}{L_M^2(x^*)^2} \tag{1.20}$$

which reduces to

$$p_M = \left(\frac{mV_M^2}{L_M^3}\right)\frac{dv^*/dt^*}{(x^*)^2} \tag{1.21}$$

Noting that $\rho = m/L^3$ and that

$$p^* = \frac{dv^*/dt^*}{(x^*)^2} \tag{1.22}$$

gives us, upon substituting into Eq. (1.21) for p_M

$$p_M = \rho V_M^2 p^* \tag{1.23}$$

A similar equation exists for the prototype, namely

$$p_P = \rho V_P^2 p^* \tag{1.24}$$

We have now derived a dimensionless variable equivalent to each dimensionalized variable in the Navier–Stokes equation. Thus we can convert the Navier–Stokes equation to a dimensionless form. Substituting the dimensionless variables for the dimensional variables in the Navier–Stokes equation for our model yields

$$\frac{\partial(V_M v^*)}{\partial(L_M/V_M)t^*} + V_M v^* \frac{\partial(V_M v^*)}{\partial(L_M x^*)} = -\frac{1}{\rho}\frac{\partial(\rho V_M^2 p^*)}{\partial(L_M x^*)} + g + \frac{\mu}{\rho}\left(\frac{\partial^2(V_M v^*)}{\partial(L_M x^*)^2}\right) \tag{1.25}$$

Grouping the constant terms in Eq. (1.25) gives us

$$\left(\frac{V_M^2}{L_M}\right)\frac{\partial v^*}{\partial t^*} + \left(\frac{V_M^2}{L_M}\right)v^*\frac{\partial v^*}{\partial x^*} = -\left(\frac{V_M^2}{L_M}\right)\frac{\partial p^*}{\partial x^*} + g + \left(\frac{\mu V_M}{\rho L_M^2}\right)\left(\frac{\partial^2 v^*}{\partial (x^*)^2}\right) \tag{1.26}$$

Dividing Eq. (1.26) by L_M/V_M^2 yields

$$\frac{\partial v^*}{\partial t^*} + v^* \frac{\partial v^*}{\partial x^*} = -\frac{\partial p^*}{\partial x^*} + \frac{gL_M}{V_M^2} + \left(\frac{\mu}{\rho L_M V_M}\right)\left(\frac{\partial^2 v^*}{\partial (x^*)^2}\right) \quad (1.27)$$

Grouping the subscripted terms provides us with

$$\frac{\partial v^*}{\partial t^*} + v^* \frac{\partial v^*}{\partial x^*} = -\frac{\partial p^*}{\partial x^*} + \left(\frac{gL.}{V^2}\right)_M + \left(\frac{\mu}{\rho L V}\right)_M \left(\frac{\partial^2 v^*}{\partial (x^*)^2}\right) \quad (1.28)$$

Equation 1.28 is the dimensionless form of the Navier–Stokes equation for our model. The dimensionless form of the Navier–Stokes equation for our prototype is

$$\frac{\partial v^*}{\partial t^*} + v^* \frac{\partial v^*}{\partial x^*} = -\frac{\partial p^*}{\partial x^*} + \left(\frac{gL.}{V^2}\right)_P + \left(\frac{\mu}{\rho L V}\right)_P \left(\frac{\partial^2 v^*}{\partial (x^*)^2}\right) \quad (1.29)$$

Inspection of Eqs. (1.28) and (1.29) shows them to be equivalent, that is, identical, if

$$\left(\frac{gL}{V^2}\right)_M = \left(\frac{gL}{V^2}\right)_P \quad (1.30)$$

and, if

$$\left(\frac{\mu}{\rho L V}\right)_M = \left(\frac{\mu}{\rho L V}\right)_P \quad (1.31)$$

Thus, if the above two conditions Eqs. (1.30 and 1.31) hold true, we have one equation describing incompressible, viscous flow in two plants of different size. Our prototype plant and our model plant will operate similarly, that is, behave identically, so long as the above conditions are met.

Each of the above conditions constitutes a dimensionless parameter. Each of them also has a name. We call first one, V^2/gL, the Froude number. It is named after William Froude, who did pioneering work in the movement of solid objects through a liquid; in particular, the movement of ships through water. We generally interpret the Froude number as

- the ratio of inertial to gravitational forces [1];
- the ratio of the kinetic energy to the gravitational potential energy of the flow [2].

The second dimensionless parameter above, namely, $\rho LV/\mu$ is the Reynolds number, named after Osborne Reynolds, who did pioneering work in fluid flow through conduits, such as pipes. We generally interpret the Reynolds number as

- the ratio of inertial to viscous forces in a fluid [1];
- the ratio of convective momentum transport to viscous momentum transport [2].

SIMILARITY

The above analysis depends on the concept of similarity, which arises from geometry. Consider two triangles, the smaller one with sides of length a, b, and c; the larger one with sides of length α, β, and γ. The two triangles are similar if their corresponding angles are equal and if [3]

$$\frac{\alpha}{a} = \frac{\beta}{b} = \frac{\gamma}{c} \tag{1.32}$$

From one model triangle, we can produce any number of other similar triangles with equal corresponding angles and measuring a, b, and c and specifying α. With this information, we can calculate β and γ. Doing so produces triangles of similar shape but different size [4]; formalizing this example yields the general concept of similarity

> *Given two scalar functions f and g, if the ratio g/f is constant at homologous points and times, then g is similar to f. The constant* $K_f = g/f$ *is the scaling factor for the function f [5].*

Homologous means that a point-to-point correspondence exists between the objects we are comparing or the function we are using to describe two objects or two events. Note that the above definition includes transient events, which requires homologous times [5].

For chemical engineers, processes are homologous when

- a pipe or vessel containing a fluid has a point-to-point correspondence for two plants of different size;
- a solid object submerged in a fluid has a point-to-point correspondence for two plants;
- the physical properties of the fluid in a conduit or vessel demonstrate point-to-point correspondence for two plants;
- the fluid composition in the conduit or vessel changes, that is, undergoes reaction, at corresponding points in the two plants [6].

We establish similarity in one of two ways: by the "shape factor" method or by the "scaling factor" method. The shape factor method involves determining all the relative locations for relevant features of two mechanisms, objects, or processes. For example, we can establish the similarity between two rectangles by measuring the equivalent sides of each rectangle, then testing the equalities

$$\frac{x_1}{y_1} = \frac{x_2}{y_2}; \quad \frac{y_1}{z_1} = \frac{y_2}{z_2}; \quad \text{and} \quad \frac{x_1}{z_1} = \frac{x_2}{z_2} \tag{1.33}$$

If the above ratios (1.33) are true, then the two rectangles are similar.

The scaling factor method involves locating a correspondence, generally a point, and establishing a scaling factor between two mechanisms, objects, or processes. For example, consider two cubes: one significantly larger than the other. A correspondence between these two cubes occurs at the center point for each cube. If from the center point

$$\frac{x_L}{x_S} = \frac{y_L}{y_S} = \frac{z_L}{z_S} = \Pi_{x,y,z} \tag{1.34}$$

where the subscript L denotes "large" and the subscript S denotes "small," then the two cubes are similar. $\Pi_{x,y,z}$ denotes a dimensionless number describing the ratios of the x, y, and z coordinates.

Four similarities are important to chemical engineers. They are

1. geometrical;
2. mechanical;
3. thermal;
4. chemical.

In general, geometric similarity means the model, with coordinates x_M, y_M, and z_M, and the prototype, with coordinates x_P, y_P, and z_P, are similar with regard to all correspondence points if

$$\frac{x_P}{x_M} = \frac{y_P}{y_M} = \frac{z_P}{z_M} = \Pi_{x,y,z} \tag{1.35}$$

$\Pi_{x,y,z}$ is the scaling factor.

Mechanical similarity comprises three subsimilarities, which are static similarity, kinematic similarity, and dynamic similarity. Static similarity demands that two geometrically similar objects have relative deformation for a constant applied stress.

Kinematic similarity means the constituent parts of a model and prototype mechanism or process in translation follow similar paths or streamlines provided the model and prototype are geometrically similar. Thus

$$\frac{v_P}{v_M} = \Pi_v \tag{1.36}$$

where v_M is the velocity of the translating model part or particle and v_P is the velocity of the translating prototype part or particle. Π_v is the velocity scaling factor.

Dynamic similarity demands that the ratio of the forces inducing acceleration be equal at corresponding locations in geometrically similar mechanisms or processes. In other words, the ratio

$$\frac{F_P}{F_M} = \Pi_F \tag{1.37}$$

holds true at every corresponding location in the two mechanisms or processes, where F_M is the force at location x_M, y_M, and z_M in the model and F_P is the force at the corresponding location x_P, y_P, and z_P in the prototype.

Thermal similarity occurs when the ratio of the temperature difference at corresponding locations of a geometrically similar mechanism or process are equal. If translation, that is, movement, occurs, then the process must also demonstrate mechanical similarity for thermal similarity to occur. Thus thermal similarity requires geometric similarity and mechanical similarity.

As chemical engineers, our major concern is the reactions occurring in the process. We want our prototype to reflect what occurs in our model. To ensure that outcome, our prototype must be chemically similar to our model. Chemical similarity demands the ratio of concentration differences at all corresponding locations in the prototype and in the model be equal. Therefore, our prototype and model must also be geometrically, mechanically, and thermally similar.

MODELS

Upscaling and downscaling involve modeling. We build models to reduce the time from ideation to commercialization and to reduce the cost of that effort. The major cost savings of modeling come from not building an inoperable, full-scale commercial plant.

There are four types of models. They are

- true models;
- adequate models;
- distorted models;
- dissimilar models.

True models involve building all significant process features to scale. Thus the model is an exact replica of the prototype, that is, of the commercial plant. We build true models in some safety investigations to determine definitely the cause of a specific, safety event. Automobile manufacturers use true models when gathering crash data about the vehicles they plan to market. For complex processes, a complete model is actually a full-scale prototype, that is, a true model [7]. While true models may provide highly accurate information, they are capital intensive, expensive to operate, and require extended time periods to build.

Adequate models predict one characteristic of the prototype accurately. In general, adequate models involve testing the dominant, controlling factor in the process. For example, porous solid–catalyzed processes are generally stagnant film diffusion rate limited or pore diffusion rate limited. If a process is so limited, then we only have to ensure the same controlling regime in our laboratory or pilot plant reactors. If we do not ensure equivalent controlling regimes in the laboratory or pilot plant reactors, then any process development or process support will be wasted effort. If we do not consider whether the commercial process is stagnant film diffusion rate limited, pore diffusion rate limited, or reaction rate limited, then we will finish our effort with an expensive scattergram of the experimental results. This situation occurs more often than we like to admit. Many porous solid–catalyzed commercial processes, that is, prototypes, are pore diffusion rate limited due to high interstitial fluid velocity through the reactor. Such high fluid velocity minimizes the boundary layer surrounding each catalyst pellet, thereby making the process pore diffusion rate limited. Unfortunately, most porous solid–catalyzed pilot plant processes, that is, models, are operated at low interstitial fluid velocities in order to minimize feed and product volumes at the research site. Both inventories can be a safety hazard if they are hydrocarbons and the product can be a disposal issue since it cannot be sold and is generally not fed into a commercial process. In such situations, stagnant film diffusion rate is the controlling regime. The result of a multiyear, multicatalyst testing effort in a stagnant film diffusion rate–limited pilot plant will be an expensive scattergram around the average value for the film diffusion rate constant. On the other hand, considerable effort can be made at the laboratory scale to ensure that catalyst testing occurs in the reaction rate–limited regime. Plots with impressive correlations result from these types of experimental programs. Unfortunately, when the best catalyst is tested in the prototype, it displays the same efficiency and productivity as the current catalyst. In such cases the prototype is either stagnant film or pore diffusion rate limited. It does not matter how reactive the catalyst is in the laboratory; in the prototype the process is incapable of keeping the catalytic active site saturated with reactant. In conclusion, the controlling regime of the model must be identical to the controlling regime of the prototype.

In distorted models we violate design conditions intentionally for one reason or another. Such distortion affects the prediction equation. Hydrologic river basin models are the most common distorted models. In these models the horizontal and vertical lengths do not have the same scaling factors. In a geometrically similar model the horizontal and vertical ratios are equal; for example,

$$\frac{{}_{P}L_{\mathrm{H}}}{{}_{M}L_{\mathrm{H}}} = \Pi_{\mathrm{H,V}} \quad \text{and} \quad \frac{{}_{P}L_{\mathrm{V}}}{{}_{M}L_{\mathrm{V}}} = \Pi_{\mathrm{H,V}} \tag{1.38}$$

where ${}_{P}L_{H}$ is the prototype characteristic horizontal length, ${}_{M}L_{H}$ is the model characteristic horizontal length, ${}_{P}L_{V}$ is the prototype characteristic vertical length, and ${}_{M}L_{V}$ is the model characteristic vertical length equivalent to ${}_{P}L_{V}$. $\Pi_{\mathrm{H,V}}$ is the scaling factor. For a distorted model

$$\frac{{}_{P}L_{\mathrm{H}}}{{}_{M}L_{\mathrm{H}}} = \Pi_{\mathrm{H}} \quad \text{and} \quad \frac{{}_{P}L_{\mathrm{V}}}{{}_{M}L_{\mathrm{V}}} = \Pi_{\mathrm{V}} \tag{1.39}$$

where Π_{H} is the horizontal scaling factor, Π_{V} is the vertical scaling factor, and $\Pi_{\mathrm{H}} \neq \Pi_{\mathrm{V}}$. It is "legal" to use distorted models, so long as we know we are doing it and we understand why we are doing it.

With regard to the process, distorted models behave in a manner similar to their prototypes; however, one dimension of the model will not be scaled equivalently to the other dimensions. Thus a distorted model may look squat or tall or broad, depending on the distortion, when compared to its prototype.

Dissimilar models comprise the fourth and last model type. Such models have no apparent resemblance and no similarity to their prototypes. These models provide information about the prototype through suitable analogies and statistical correlations.

DIMENSIONAL ANALYSIS AND MODELS

Two chemical processes are similar if their dimensionless geometric ratios are equal and if their dimensionless mechanical, thermal, and chemical parameters are equal. When corresponding parameters are equal, then the prototype and model are similar. Symbolically

$$\begin{aligned}
\Pi_{\mathrm{M}}^{\mathrm{Geometric}} &= \Pi_{\mathrm{P}}^{\mathrm{Geometric}} \\
\Pi_{\mathrm{M}}^{\mathrm{Static}} &= \Pi_{\mathrm{P}}^{\mathrm{Static}} \\
\Pi_{\mathrm{M}}^{\mathrm{Kinematic}} &= \Pi_{\mathrm{P}}^{\mathrm{Kinematic}} \\
\Pi_{\mathrm{M}}^{\mathrm{Dynamic}} &= \Pi_{\mathrm{P}}^{\mathrm{Dynamic}} \\
\Pi_{\mathrm{M}}^{\mathrm{Thermal}} &= \Pi_{\mathrm{P}}^{\mathrm{Thermal}} \\
\Pi_{\mathrm{M}}^{\mathrm{Chemical}} &= \Pi_{\mathrm{P}}^{\mathrm{Chemical}}
\end{aligned} \tag{1.40}$$

where the Π subscript M designates the model and the Π subscript P designates the prototype.

The most general equation for a prototype is

$$\Pi_1^{\mathrm{P}} = f(\Pi_2^{\mathrm{P}}, \Pi_3^{\mathrm{P}}, \ldots, \Pi_n^{\mathrm{P}}) \tag{1.41}$$

where the subscript numeral identifies a dimensionless parameter and superscript P indicates prototype. This equation applies to all mechanisms or processes that are comprised of the same dimensional variables. Thus it applies to any model of the same mechanism or process, which means we can write a similar equation for that model

$$\Pi_1^{\mathrm{M}} = f(\Pi_2^{\mathrm{M}}, \Pi_3^{\mathrm{M}}, \ldots, \Pi_n^{\mathrm{M}}) \tag{1.42}$$

Dividing Eq. (1.41) by Eq. (1.42) gives us

$$\frac{\Pi_1^{\mathrm{P}}}{\Pi_1^{\mathrm{M}}} = \frac{f(\Pi_2^{\mathrm{P}}, \Pi_3^{\mathrm{P}}, \ldots, \Pi_n^{\mathrm{P}})}{f(\Pi_2^{\mathrm{M}}, \Pi_3^{\mathrm{M}}, \ldots, \Pi_n^{\mathrm{M}})} \tag{1.43}$$

Note that if $\Pi_2^{\mathrm{P}} = \Pi_2^{\mathrm{M}}$and $\Pi_3^{\mathrm{P}} = \Pi_3^{\mathrm{M}}$ and so on, thus

$$\frac{\Pi_1^{\mathrm{P}}}{\Pi_1^{\mathrm{M}}} = 1 \tag{1.44}$$

which is the condition for predicting prototype behavior from model behavior. The conditions

$$\begin{aligned} \Pi_2^{\mathrm{P}} &= \Pi_2^{\mathrm{M}} \\ \Pi_3^{\mathrm{P}} &= \Pi_3^{\mathrm{M}} \\ &\vdots \\ \Pi_n^{\mathrm{P}} &= \Pi_n^{\mathrm{M}} \end{aligned} \tag{1.45}$$

constitute the design specifications for the prototype from the model or the model from the prototype, depending on whether we are upscaling or downscaling. If all these conditions are met, then we have a valid model. If the above conditions hold for the controlling regime of the model and the prototype, then we have an adequate model. If most of the above conditions hold true, then we have a distorted model that requires a correlation to relate Π_1^{P} and Π_1^{M}; in other words, we need an additional function such that

$$\Pi_1^{\mathrm{P}} = f(\text{correlation})\Pi_1^{\mathrm{M}} \tag{1.46}$$

If none of the above conditions holds true, then we have an dissimilar model.

We generally do not build true models in the chemical processing industry (CPI). A true model of a chemical process implies building a commercial-sized plant, which is far too costly and time consuming for an organization to undertake. Most models in the CPI are adequate or distorted models. Of these two types, adequate models are the better since they model the controlling regime of the process. Distorted models are more difficult to use because we have to determine the correlation between the distorted model and the prototype.

SUMMARY

In this chapter, we discussed the concepts of models and prototypes and presented the criteria for scaling a chemical process from a model to a prototype or from a commercial plant to a pilot plant or a laboratory. We also introduced the concepts of "upscaling" and "downscaling" in this chapter.

REFERENCES

[1] R. Johnstone, M. Thring, Pilot Plants, Models, and Scale-up Methods In Chemical Engineering, McGraw-Hill Book Company, Inc., New York, NY, 1957, p. 51.

[2] W. Krantz, Scaling Analysis in Modeling Transport and Reaction Processes: A Systematic Approach to Model Building and the Art of Approximation, John Wiley and Sons, Inc, New York, NY, 2007, p. 513.

[3] W. Duncan, Physical Similarity and Dimensional Analysis: An Elementary Text, Edward Arnold & Co., London, UK, 1953, p. 1.

[4] V. Skoglund, Similitude: Theory and Applications, International Textbook Company, Scranton, PA, 1967, p. 13.

[5] H. Langhaar, Dimensional Analysis and Theory of Models, John Wiley and Sons, Inc, New York, NY, 1951, p. 68.

[6] R. Johnstone, M. Thring, Pilot Plants, Models, and Scale-up Methods in Chemical Engineering, McGraw-Hill Book Company, Inc, New York, NY, 1957, p. 74.

[7] V. Skoglund, Similitude: Theory and Applications, International Textbook Company, Scranton, PA, 1967, pp. 74–75.

CHAPTER 2

Dimensional Analysis

INTRODUCTION

To ensure that a prototype chemical process behaves similarly to its model chemical process, we must establish the relationships

$$\begin{aligned}
\Pi_M^{Geometric} &= \Pi_P^{Geometric} \\
\Pi_M^{Static} &= \Pi_P^{Static} \\
\Pi_M^{Kinematic} &= \Pi_P^{Kinematic} \\
\Pi_M^{Dynamic} &= \Pi_P^{Dynamic} \\
\Pi_M^{Thermal} &= \Pi_P^{Thermal} \\
\Pi_M^{Chemical} &= \Pi_P^{Chemical}
\end{aligned} \tag{2.1}$$

where each Π_M^i and Π_P^i is a dimensionless parameter for the model M and the prototype P. The superscript i designates the required similarity.

We can determine each of the above dimensionless parameters by establishing characteristic lengths, velocities, forces, etc., for the conservation equations governing the chemical process, then manipulating them as we did in chapter "Introduction" to obtain a set of dimensionless parameters. However, this procedure is time consuming since we have to define a characteristic length, velocity, force, etc., that is valid for each conservation equation involved in describing the chemical process. It is much preferable to develop an algebraic procedure for determining dimensionless parameters.

In this chapter, we develop an algebraic procedure, using linear algebra, that is, matrices, for determining dimensionless parameters. This linear algebra procedure is not new—it was proposed and developed by several engineers during the 1950s. At that time, however, we solved matrices by "hand," which is a slow process fraught with potential error. Aeronautical, civil, and mechanical engineers used a matrix-based procedure to generate dimensionless parameters during the 1950s and 1960s for mechanical processes. These mechanical processes

Scaling Chemical Processes.
© 2016 Elsevier Inc. All rights reserved.

did not generate large matrices. Chemical engineers did not use the matrix procedure during the same period because the resulting matrices were too large to manipulate by "hand." However, with the advent of the personal computer and user-friendly software, we can now manipulate and solve large matrices quickly. And the necessary software for solving large matrices is available for free use in the Internet.

This chapter presents the foundation for determining dimensionless parameters easily and quickly via a matrix procedure. But, before we can develop this matrix procedure, we must first distinguish the difference between dimension and unit.

PHYSICAL CONCEPT, PHYSICAL QUANTITY, AND DIMENSION

Language is a precision instrument. Unfortunately, we tend to use it imprecisely. For example, we use "dimensions" and "units" interchangeably when, in fact, they denote entirely different concepts.

Dimension arises from our desire to quantify our perceptions. We describe a sensory perception with a physical concept. Force we experience as impact. If a grocery cart bumps a car in the parking lot, the car may suffer a scratch or small dent. If a car collides with another car at a city street intersection, the damage to both cars will be significant. If an 18-wheel truck collides with a car on an interstate or motorway, the result will be catastrophic. From this information, we know that force, or impact, involves mass and acceleration. Thus, to quantify force, we need to specify the dimensions mass, length, and time. When we make such a specification, we convert the physical concept of force into the physical quantity of force. Thus to quantify a physical concept, we first determine a descriptor that best characterizes it, that produces a valid physical quantity. Therefore a physical quantity represents a qualitative description of a physical concept. In other words, a physical concept attains meaning only if its descriptors can be measured. Those descriptors are dimensions.

A dimensional system comprises the fewest dimensions necessary to quantify a particular feature of Nature. The necessary dimensions form a basis set with which to describe our perceptions of Nature. The dimension basis set for the Le Systeme International d'Unites (SI units) is: length [L], mass [M], time [T], thermodynamic temperature [θ], amount of substance [N], electric current [A], and luminous intensity [CD].

James Clerk Maxwell suggested, not clearly, that we designate the physical quantity of a physical concept with brackets. For example, consider the physical concept of distance: the physical quantity of distance is $[L^1]$; for area, the physical quantity is $[L^2]$; and, for volume, the physical quantity is $[L^3]$. The power indices on L actually designate "dimension" [1]. Thus, the index 1, or unit power, designates lines; the index 2, or square power, designates planes; that is, areas; and, the index 3 designates spaces; that is, volumes.

Dimensions come in two varieties: fundamental and derived. Fundamental dimensions form a basic set of quantifiers for describing a physical concept. Derived dimensions arise from our study of Nature: many observations of Nature require us to combine fundamental dimensions to quantify them. Chasing mastodons across the prairie may have suggested the physical concept of speed to one of our ancestors. Speed, or velocity, is described by the physical quantity of [L]/[T] or $[LT^{-1}]$. $[LT^{-1}]$ is a derived dimension. Using the fundamental dimensions [LMT], we derive force [F] via Newton's second law, namely

$$\begin{aligned} &F = ma \\ &\mathrm{F[=]ML/T^2 = MLT^{-2}} \end{aligned} \tag{2.2}$$

where [=] implies our equation uses dimension notation rather than mathematical notation. Using the fundamental dimensions [FLT], we derive mass [M] using the same law; in other words

$$\begin{aligned} &F = ma \\ &F/a = m \\ &\mathrm{M[=]FT^2/L = L^{-1}FT^2} \end{aligned} \tag{2.3}$$

We choose the LMT fundamental dimension set for situations involving dynamics and we choose the LFT fundamental dimension set for situations involving statics. Table 2.1 shows various fundamental dimension sets and their derived dimensions.

Note that neither fundamental dimensions nor derived dimensions involve magnitudes. Dimension describes the nature of a physical concept without introducing magnitude [2].

Dimension associates a rule or procedure; that is, a standard measuring procedure, with a physical concept. We call the standard measuring procedure a "descriptor": it describes what needs measuring to transform a physical concept into a physical quantity.

Table 2.1 Various Fundamental Dimensions

Physical Quantity	Fundamental Dimensions					
	LMT	LFT	LET	LMTθ	LFTθ	LQTθ
Angle	L^0	L^0	L^0	L^0	L^0	L^0
Line	L^1	L^1	L^1	L^1	L^1	L^1
Area	L^2	L^2	L^2	L^2	L^2	L^2
Volume	L^3	L^3	L^3	L^3	L^3	L^3
Velocity	LT^{-1}	LT^{-1}	LT^{-1}	LT^{-1}	LT^{-1}	LT^{-1}
Acceleration	LT^{-2}	LT^{-2}	LT^{-2}	LT^{-2}	LT^{-2}	LT^{-2}
Force, weight	LMT^{-2}	F	$L^{-1}E$	LMT^{-2}	F	$L^{-1}E$
Density	$L^{-3}M$	$L^{-3}F$	$L^{-4}E$	$L^{-3}M$	$L^{-3}F$	$L^{-4}E$
Energy	L^2MT^{-2}	LF	E	L^2MT^{-2}	LF	E
Power	L^2MT^{-3}	LFT^{-1}	ET^{-1}	L^2MT^{-3}	LFT^{-1}	ET^{-1}
Mass	M	$L^{-1}FT^2$	$L^{-2}ET^2$	M	$L^{-1}FT^2$	$L^{-2}ET^2$
Entropy				$L^2MT^{-2}\theta^{-1}$	$LF\theta^{-1}$	$E\theta^{-1}$
Thermal conductivity				$LMT^{-3}\theta^{-1}$	$T^{-1}F\theta^{-1}$	$L^{-1}T^{-1}E\theta^{-1}$

EQUATIONS AND PHYSICAL MAGNITUDE

Equations come in two varieties: mathematical and physical. Mathematical equations involve numbers that have no physical content; that is, they involve pure numbers. We explore the relationships between pure numbers using the rules of mathematics. We learn a fair number of these relationships during our mathematical preparation for an engineering career.

Scientists and engineers use physical equations. Physical equations are developed from experimental data and observation. They balance one set of physical magnitudes against another set of physical magnitudes via the equality sign of mathematics. The law for the conservation of energy is a good example of a physical equation. It was developed during the mid-19th century through the effort of many scientists and engineers. For a flowing fluid, the physical concept for the mechanical conservation of energy is

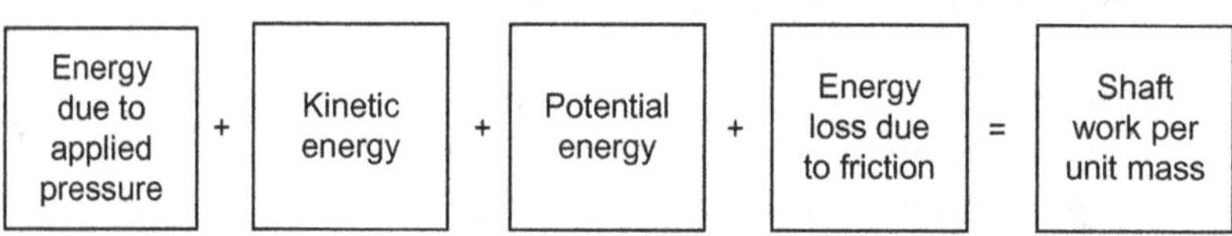

A physical quantity tells us what dimensions we require for describing a given physical concept. To convert a physical quantity into a physical magnitude, we require a system of units. If we denote the physical concept as α and its descriptor as $[\psi]$, where ψ denotes a fundamental or derived dimension, then we can denote the physical quantity as $\alpha[\psi]$. A physical magnitude is any physical quantity to which we can affix a numerical value. Thus physical magnitudes are [3]

$$\alpha[\psi] = \{\text{Numerical value}\} \cdot \{\text{Standard unit of measure}\} \pm \text{Error} \quad (2.4)$$

Note that we can establish a physical concept for taste and bouquet and physical quantities for both, with dimension $[\tau]$ for taste and dimension $[\sigma]$ for smell. However, we cannot convert these physical quantities into physical magnitudes because we are unable to establish a standard unit of measure for either, which means we are unable to assign a numerical value to the physical quantity [4].

For example, let us consider a piece of rope. Its physical concept is the distance from one rope end to the other rope end. We convert this physical concept to a physical quantity by specifying a dimension. That dimension is length [L]. Specifying a dimension means establishing a procedure for quantifying the physical concept. In this case, we lay the piece of rope on the ground and draw it taut, then we pin each end to the ground. Next, we choose a straight stick as our unit and cut several more straight sticks of the same length. The miniscule difference in each stick length constitutes our measurement error. To convert the physical quantity to a physical magnitude, we take a straight stick and align one end of it adjacent to a fixed end of the rope. Then, we lay a second stick end to end against the first stick and along the rope. We continue adding similar sticks until we reach the rope's second fixed end. The physical magnitude of the rope is

$$(\text{Number of sticks}) \cdot (1 \text{ stick}) \pm \text{Measuring error} \quad (2.5)$$

If we laid six sticks along the rope, then the physical magnitude of the rope is $6 \cdot (1 \text{ stick})$. If our sticks are the length of our foot, then

$$1 \text{ stick} = 1 \text{ foot} \quad (2.6)$$

and the physical magnitude of the rope would be $6 \cdot (1 \text{ foot})$ or 6 feet $\pm$ measurement error.

SYSTEMS OF UNITS

A unit is a specified quantity for a given dimension; thus it allows us to create physical magnitudes. We can establish, in any fashion, a unit to determine the magnitude of a specified dimension, just so long as we follow the approved procedure for using that unit. For example, the scientific community established, in 1889 and used until 1960, the International Meter, defined as the length between two engraved lines on a bar of 90% platinum, 10% iridium, as the standard unit of length. After 1960, the scientific community defined the standard unit of length as 1,650,763.73 vacuum wavelengths of 6058 Å light emitted by a Krypton-86 discharge tube. We call this length the optical meter. The accuracy for the platinum–iridium bar is $1/(1 \times 10^6)$, whereas the optical meter has an accuracy of $1/(1 \times 10^8)$ [5]. The scientific community redefined the optical meter in 1983. It is now the distance traveled by light in vacuum in 1/299,792,458 s. ASTM International provides a library of procedures for specifying standard units and for using standard units for measurement.

Units can be extensive or intensive. Extensive units sum to become larger. Length has an extensive unit; that is, 5 feet of rope may be one piece or it may be $5 \cdot 1$ foot rope segments joined together. Mass has an extensive unit. On the other hand, intensive units do not sum to become larger; that is, temperature, density, and viscosity have intensive units [6–8].

Systems of standard units, or simply, systems of units are ubiquitous in science and engineering. Historically, each scientific and engineering discipline developed a system of units that best matched their needs. Each of these systems of units is valid, so long as their practitioners use them properly and correctly. This chaotic democracy requires us to know the foundations of each system of units and it requires an inordinate number of conversion factors [9]. The need for these conversion factors is disappearing as more of the global community adopts the International System (SI) of Units. However, the United States and some other countries still use the English Engineering system of units and the American Engineering system of units. Also, much of the scientific and engineering literature, especially prior to the mid-1960s, uses systems of units other than metric or SI units.

Any number of unit systems can be devised, so long as the fundamental dimensions upon which they are founded are independent of each other. The simplest such unit systems are based on three fundamental dimensions. These fundamental dimensions are Length, Mass, and Time, or Length, Force, and Time. Newton's second law links these fundamental dimensions, namely, for Length, Mass, and Time

$$F = ma = [\mathrm{LMT}^{-2}] \tag{2.7}$$

and for Length, Force, and Time

$$m = \frac{F}{a} = [\mathrm{L}^{-1}\mathrm{FT}^{2}] \tag{2.8}$$

In the former system of units, 1 kilogram (kg) of mass accelerated at 1 meter per second squared ($\mathrm{m/s^2}$) experiences a force of 9.81 kilogram · meter/time squared ($\mathrm{kg \cdot m/s^2}$). In the latter system of units, 9.81 $\mathrm{kg \cdot m/s^2}$ of force imparts an acceleration of 1 $\mathrm{m/s^2}$ to 1 kg of mass. Neither of these systems of units requires a dimensional constant. We call such systems of units "absolute" [10]. We refer to absolute systems of units based on Length, Mass, and Time as dynamic systems of units [11]. Physicists and engineers designing mechanisms with moving parts use dynamic systems of units.

We call systems of units based on Length, Force, and Time static systems of units. We sometimes call them gravitational systems of units [11]. Structural engineers, civil engineers, and engineers designing stationary equipments prefer to use static systems of units. Thus, in dynamic and gravitational systems of units, Newton's second law does not require a constant; it is, simply

$$F = ma \tag{2.9}$$

Before the metric system of units gained wide popularity, Great Britain used foot, pound, and second as its system of units. In this system of units, 1 pound force imparts 32 foot per second squared acceleration to 1 pound mass. This system of units was widely used until the mid-1960s. At the same time, mechanical engineers in Great Britain wanted to use the pound to define force, whereas chemists, and later chemical engineers, wanted to use the pound for mass [9]. To accommodate both groups, engineers and scientists started using a system of units with four fundamental dimensions, those dimensions being Length, Mass, Force, and Time.

We call systems of units using fundamental dimensions of Length, Mass, Force, and Time "engineering systems of units". An enormous amount of confusion exists in the chemical processing industry because of these systems of units. The confusion arises because we give Force and Mass the same unit identification, namely, kilogram or pound. Also, when publishing results, scientists and engineers tend not to specify which system of units they are using in their memo, paper, or report. Thus we, the readers, must determine whether the author is using an engineering, gravitational, or dynamic system of units. Adoption of the SI system of units solves this problem for future scientists and engineers; however, much of our technical information is historic. Thus when we read a published paper or report, we must determine which system of units the author used to correctly implement his or her data and conclusions.

Historically the four most common engineering systems of units are

- Old English Engineering [12];
- New English Engineering;
- English Gravitational (Static);
- English Dynamic (Absolute).

Table 2.2 presents these systems of units, as well as the fundamental dimensions defining each unit system. While we may apply the above names as though they were commonly used, they were not and are not. Each system of units has been identified differently when authors have written about units. For example, both the Old English and New English Engineering systems of units have been identified as the "American Engineering" system of units. Also, some authors have identified the English Gravitational system of units as the "British Engineering" system of units [9]. We identify the above systems of

Table 2.2 English Engineering Fundamental Dimensions and Units

Dimensions	Symbols	Old English Units	New English Units	English Gravitational Units	English Dynamic Units
Length	L	Foot	Foot	Foot	Foot
Mass	M	Pound mass (lb)	Pound mass (lb_M)	Slug	Pound (lb)
Time	T	Second or hour	Second or hour	Second	Second
Force	F	Pound force (lb)	Pound force (lb_F)	Pound	Poundal
Thermodynamic temperature	Θ	°F	°C	°R	°C

units collectively as "mechanical" systems of units because all of mechanics can be described using them.

The Old English Engineering system of units uses pound (lb) for force, pound (lb) for mass, foot (ft) for length, and second or hour (s or hr) for time. Note that lb for force and lb for mass are not stipulated in this system of units. You have to be an astute, careful reader to distinguish which pound an author is using. This system of units has caused much confusion among scientists and engineers.

To alleviate that confusion, engineers proposed the New English Engineering unit system, in which pound force (lb_F) and pound mass (lb_M) are separately identified. This notation reduces the confusion inherent in identifying two units with the same symbol.

This identification of lb_F and lb_M brings us to the underlying cause of confusion concerning engineering systems of units; namely, force is defined and not derived. Newton's second law states that

$$F = ma$$

In other words, we have two independent variables and one dependent variable. We can specify the dimensions for any two variables, thereby allowing the third variable to become the dependent variable whose dimensions are derived from the dimensions of the other two variables. However, when we specify the dimensions of all three variables, we overspecify the equation. Thus in the Old English Engineering system of units Newton's second law has units of

$$\mathrm{lb} = \mathrm{lb}(\mathrm{ft}/\mathrm{s}^2) \tag{2.10}$$

which is highly confusing. In the New English Engineering system of units Newton's second law has units of

$$\mathrm{lb_F} = \mathrm{lb_M}(\mathrm{ft}/\mathrm{s}^2) \tag{2.11}$$

From Eqs. (2.10) and (2.11), we immediately see that we need a dimensional constant to achieve dimensional homogeneity. Dimensional homogeneity means the dimensions on both sides of an equality sign are the same. That dimensional constant is the gravitational constant g_O and g_C, respectively.

Consider the gravitational constant g_C: what is it? We define Newton's second law by measuring the acceleration that one pound force imparts to one pound mass during free fall. We stipulate that free fall occurs at 45 degree latitude and at sea level. Free fall acceleration thus defined is 32 ft/s^2. Newton's second law, therefore, becomes

$$1\ \text{lb}_\text{F} = 1\ \text{lb}_\text{M}(32\ \text{ft/s}^2) \tag{2.12}$$

To dimensionally balance Eq. (2.12), we see that g_C must be 32 lb_Mft/lb_F s^2. Newton's second law thus becomes

$$(32\ \text{lb}_\text{M}\text{ft/lb}_\text{F}\text{s}^2)(1\ \text{lb}_\text{F}) = (1\ \text{lb}_\text{M})(32\ \text{ft/s}^2) \tag{2.13}$$

In mathematical notation, we write Eq. (2.13) as

$$g_C F = ma \tag{2.14}$$

What about g_O? In the Old English Engineering system of units we write Newton's Second Law as

$$\text{lb} = \text{lb}(32\ \text{ft/s}^2) \tag{2.15}$$

g_O is measured using the same stipulations used to measure g_C. It is 32 ft/s^2. Thus for the Old English Engineering system of units Newton's second law is

$$(32\ \text{ft/s}^2)\text{lb} = \text{lb}(32\ \text{ft/s}^2) \tag{2.16}$$

and, in mathematical notation it is

$$g_O F = ma \tag{2.17}$$

In the English Gravitational system of units, the fundamental dimensions are length in feet (ft), force (lb or lb_F), and time in seconds or hours (s or hr). In this system of units, a one pound force imparts an acceleration of 32 ft/s^2 to a mass of one "slug." Newton's second law is then

$$1\ \text{lb} = (1\ \text{slug})\ (32\ \text{ft/s}^2) \tag{2.18}$$

Thus

$$1\ \text{slug} = (1/32)\ \text{lb s}^2/\text{ft} \tag{2.19}$$

In the English Dynamic system of units, one pound of mass accelerated at 32 ft/s^2 experiences a force of one poundal. From Newton's second law, one poundal is

$$1 \text{ poundal} = (1 \text{ lb})\,(32 \text{ ft/s}^2) = 32 \text{ lb ft/s}^2 \tag{2.20}$$

Thus the state of confusion concerning units.

DEVELOPING DIMENSIONAL ANALYSIS

Since physical equations contain physical magnitudes, they must by necessity contain physical content. They contain physical content because physical magnitudes arise from physical quantities, which in turn arise from our perceptions. Therefore, when we write a physical equation, we are, in essence, writing an equation that balances physical quantities $\alpha[\Psi]$ through the use of an equality sign. Thus, we arrive at the first "axiom" of Dimensional Analysis.

Axiom 1

The numerical equality of a physical equation exists only when the physical magnitudes of that particular physical equation are similar; that is, have the same units, which means the dimensions of the underlying physical quantities $\alpha[\Psi]$ are similar [13].

In other words, a valid physical equation is dimensionally homogeneous; that is, all its terms have the same dimensions and units. Joseph Fourier was the first to explicitly state this concept, which he did in the 1822 edition of his book on heat flow.

All engineers and scientists learn this axiom upon their introduction to the study of Nature. We are told, upon writing and solving our first physical equation, that the individual terms of the given physical equation must have the same dimensions; therefore, the units of each individual term in the physical equation will be the same. We are also told that the dimensions and units of our calculated result must agree with the dimensions and units of the individual terms of the physical equation. For example, consider the physical equation

$$W = X - Y + Z \tag{2.21}$$

We can only calculate W if X, Y, and Z have the same dimensions and units. If X, Y, and Z each represents a physical magnitude of apples, then we can add and subtract them to obtain W, which will be a physical magnitude of apples. If X and Z have apple dimension and Y has orange dimension, then the above expression (2.21) ceases to be a physical equation; it becomes meaningless from an engineering or a scientific viewpoint. Nonhomogeneous expressions do not contain physical information, thus they are not physical equations. The classic example of a nonhomogeneous expression is

$$s + v = \frac{1}{2}at^2 + at \tag{2.22}$$

where s is distance [L]; v is velocity $[LT^{-1}]$; a is acceleration $[LT^{-2}]$; and t is time [T] [14,15]. Writing this expression in dimensional terms gives us

$$[L] + [LT^{-1}] = [LT^{-2}][T^2] + [LT^{-2}][T] \tag{2.23}$$

which yields, upon simplification

$$[L] + [LT^{-1}] = [L] + [LT^{-1}] \tag{2.24}$$

Expression (2.24) contains information, but that information does not describe a relationship between the left and right side of the equality sign. No such relationship exists because the dimensions of the individual terms of the expression are mismatched. We frequently encounter nonhomogeneous expressions during our professional careers. Such expressions generally correlate, statistically, a product property to a process variable. In other words, the correlation describes a coincidence, not a cause and effect. Many such correlations exist in the polymer industry. Unfortunately, each such correlation is valid only for a given product from a particular production plant, which means the correlation possesses no physical information for another product or a different production plant.

We classify homogeneous physical equations as "restricted" and as "general." An example of a restricted equation is

$$s = (16.1)t^2 \tag{2.25}$$

which describes the distance s [L] traversed by a free-falling object in time t [T]. Dimensionally the above expression (2.25) is

$$[L] = [T^2] \tag{2.26}$$

which makes it nonhomogeneous. However, we know, in certain situations, that it contains valid physical information. For this expression to be true, the coefficient 16.1 must have dimensions $[LT^{-2}]$. It, therefore, is not unreasonable for us to assume

$$16.1 = \frac{1}{2} g_O \tag{2.27}$$

where g_O is 32.2 ft/s^2 in the Old English Engineering system of units. Hence

$$s = (16.1)t^2$$

is a valid physical equation so long as the coefficient is a dimensional constant with Old English Engineering units. If this condition is true, the above expression (2.25) becomes a restricted homogeneous physical equation. However, the above expression (2.25) is not a physical equation if we use the SI system of units.

Now, consider Newton's second law

$$F = ma$$

It is an example of a general homogeneous physical equation since the dimensions on either side of the equality sign are $[LMT^{-2}]$. Its physical magnitudes can be expressed using any consistent system of units. Note that a general homogeneous physical equation does not contain a dimensional constant [16].

Consider our first ancestor who described to his fellow cave mates the concept of length and how to make a spear. To demonstrate how long to make a spear, he placed a straight, trimmed sapling on the cave floor and ensured that its larger end touched the cave wall. He then took his club and laid it beside the future spear, again ensuring that the end of the club touched the cave wall. Our ancestor then upended the club and walked it along the length of the future spear, counting each upending, until he reached its tip. Thus our ancestor found the length of the future spear relative to the length of his club. Symbolically, he found

$$L_{Spear} = \alpha L_{Club} \tag{2.28}$$

where α is the number of times he upended the club from spear butt to spear tip. α is a pure number that we can manipulate with the logic and rules of mathematics. Note that L_{Spear} and L_{Club} are physical concepts; that is, they are symbols and are not subject to the logic and rules of mathematics. Looking at his fellow cave conferees, our ancestor realizes that clubs come in a variety of lengths. So, he decides to step-off the length of the future spear using his feet since most people have similar foot lengths. He, therefore, backed against the cave wall and began stepping heel-to-toe along the length of the future spear, then he did the same along the length of his club. He found that

$$L_{Spear} = \beta L_{foot} \tag{2.29}$$

and

$$L_{Club} = \gamma L_{foot} \tag{2.30}$$

Scratching his head, our ancestor realizes that the ratio of the future spear length to club length equals a pure number, namely

$$\frac{L_{Spear}}{L_{Club}} = \alpha \tag{2.31}$$

He also realized the same is true for his second measurement, hence

$$\frac{L_{Spear}}{L_{Club}} = \frac{\beta\, L_{foot}}{\gamma\, L_{foot}} \tag{2.32}$$

But, the ratio of L_{Foot} is constant and can be deleted from this ratio. Thus

$$\frac{L_{Spear}}{L_{Club}} = \frac{\beta}{\gamma} \tag{2.33}$$

Equating the two ratios, our ancestor obtained

$$\frac{L_{Spear}}{L_{Club}} = \frac{\beta}{\gamma} = \alpha \tag{2.34}$$

Since α, β, and γ are pure numbers, our ancestor realized that the ratio of two physical quantities, in this case L_{Spear} and L_{Club}, is equal to the ratio of the numbers of units used to measure them, regardless of the system of units used to measure them [17]. In other words the ratio of physical magnitudes of similar dimension is independent of the system of units. Thus the ratio of physical magnitudes possesses an absolute significance independent of the system of units used to measure the corresponding physical quantity [18].

Note that the above result makes it inherent that physical magnitude is inversely proportional to the size of the unit used, which is due to the linearity of our fundamental dimensions [19]. This result brings us to the second axiom of Dimensional Analysis, which states

Axiom 2

The ratio of physical magnitudes of two like physical quantities $\alpha[\Psi]$ is independent of the system of units used to quantify them, so long as the numerator and denominator of the ratio use the same system of units [13].

For example, our ancestor, the spear maker, owns a garden plot that is 50 feet by 100 feet. He wanted to sell it for quite some time and has finally found a buyer for it. This buyer, unfortunately, lives in the neighboring kingdom where they measure length in "rods." The buyer has no idea what is a foot and our ancestor has no idea what is a rod. Therefore, the buyer brings his measuring rod to our ancestor's garden plot and finds it to be 3 rods by 6 rods.

The ratio of the length to breadth ratio of our ancestor's garden plot is

$$\frac{100\text{ feet}}{50\text{ feet}} = 2 \tag{2.35}$$

and in rods the ratio is

$$\frac{6\text{ rods}}{3\text{ rods}} = 2 \tag{2.36}$$

as per Axiom 2. Note that the resulting ratios are dimensionless. Dividing one ratio (2.35) by the other (2.36) yields

$$\frac{100\text{ feet}/50\text{ feet}}{6\text{ rods}/3\text{ rods}} = \frac{2}{2} = 1 \tag{2.37}$$

Thus we can equate the two ratios (2.35 and 2.36), namely

$$\frac{100\text{ feet}}{50\text{ feet}} = \frac{6\text{ rods}}{3\text{ rods}} \tag{2.38}$$

which means that, within a given set of fundamental dimensions, all systems of units are equivalent. In other words, there is no distinguished or preferred system of units for a given set of fundamental dimensions.

We can also demonstrate Axiom 2 using a common engineering ratio. Consider the Reynolds number for fluid flowing in a pipe, which is defined as

$$Re = \frac{\rho D v}{\mu} \tag{2.39}$$

where ρ is fluid density $[ML^{-3}]$; D is the pipe's diameter [L]; v is fluid velocity $[LT^{-1}]$; and μ is fluid viscosity $[L^{-1}MT^{-1}]$. In the English Engineering system of units, the density of water at 20°C is 62.3 lb_M/ft [3] and its viscosity is 2.36 lb_M/ft · hr or 0.000655 lb_M/ft · s. If the pipe's diameter is 1 foot and the water is flowing at 100 ft/s, then the Reynolds number is

$$Re = \frac{(62.3\ lb_M/ft^3)(1\ ft)(100\ ft/s)}{0.000655\ lb_M/ft \cdot s} = \frac{6230\ lb_M/ft \cdot s}{0.000655\ lb_M/ft \cdot s} = 9.5 \times 10^6 \tag{2.40}$$

In the SI system of units, water density at 20°C is 998 kg/m^3 and its viscosity is 0.000977 kg/m · s. The equivalent pipe diameter is 0.305 m and the equivalent water flow rate is 30.5 m/s. The Reynolds number is, then

$$Re = \frac{(998\ kg/m^3)(0.305\ m)(30.5\ m/s)}{0.000977\ kg/m \cdot s} = \frac{9284\ kg/m \cdot s}{0.000977\ kg/m \cdot s} = 9.5 \times 10^6 \tag{2.41}$$

Equating (2.40) and (2.41), we get

$$\frac{(998\ kg/m^3)(0.305\ m)(30.5\ m/s)}{0.000977\ kg/m \cdot s} = \frac{(62.3\ lb_M/ft^3)(1\ ft)(100\ ft/s)}{0.000655\ lb_M/ft \cdot s} = 9.5 \times 10^6 \tag{2.42}$$

which shows that the English Engineering system of units is equivalent to the SI system of units. This result again suggests that no distinguished or preferred system of units exists for any set of dimensions.

DIMENSION AS A POWER LAW

We can generalize this suggestion by considering a physical concept α that we want to quantify. Our first step is to choose a set of fundamental dimensions $[\Psi]$ that will quantify α. For example, let us choose Length, Mass, and Time (LMT) as our fundamental dimension set. We next select the system of units we will use to determine the physical magnitude of α. Since there are many such systems of units, let us choose $L_1M_1T_1$ as our system of units. Thus

$$\alpha[LMT] = \Phi(L_1, M_1, T_1) \tag{2.43}$$

where α represents a physical concept and $[\Psi]$ represents the fundamental dimensions quantifying α. $\Phi(L_1,M_1,T_1)$ represents the function determining the physical magnitude in the chosen system of units. We could have chosen a different system of units, which we identify as $L_2M_2T_2$. Note that $L_1M_1T_1$ and $L_2M_2T_2$ are related by a constant, β, which we have identified as a "conversion factor." Mathematically the two systems of units are related as

$$\beta = \frac{\Phi(L_2, M_2, T_2)}{\Phi(L_1, M_1, T_1)} \tag{2.44}$$

Converting our physical quantity from the $L_1M_1T_1$ system of units to the $L_2M_2T_2$ system of units involves substituting $\Phi(L_2, M_2, T_2)/\beta$ for $\Phi(L_1,M_1,T_1)$; thus

$$\alpha[LMT] = \frac{\Phi(L_2, M_2, T_2)}{\beta} \tag{2.45}$$

Now, consider a third system of units designated $L_3M_3T_3$. Converting our physical quantity from the $L_1M_1T_1$ system of units to the $L_3M_3T_3$ system of units involves yet another conversion factor

$$\gamma = \frac{\Phi(L_3, M_3, T_3)}{\Phi(L_1, M_1, T_1)} \tag{2.46}$$

which upon substituting into

$$\alpha[LMT] = \Phi(L_1, M_1, T_1)$$

yields

$$\alpha[LMT] = \frac{\Phi(L_3, M_3, T_3)}{\gamma} \tag{2.47}$$

Dividing conversion (2.47) by conversion (2.45) gives

$$\frac{\alpha[\mathrm{LMT}]}{\alpha[\mathrm{LMT}]} = \frac{\Phi(L_3, M_3, T_3)/\gamma}{\Phi(L_2, M_2, T_2)/\beta} = \frac{\beta\Phi(L_3, M_3, T_3)}{\gamma\Phi(L_2, M_2, T_2)} = 1 \tag{2.48}$$

Thus

$$\frac{\gamma}{\beta} = \frac{\Phi(L_3, M_3, T_3)}{\Phi(L_2, M_2, T_2)} \tag{2.49}$$

Note that we could have done each of these conversions via a different route; namely, we could have converted each unit individually. Let us return to the conversion

$$\beta = \frac{\Phi(L_2, M_2, T_2)}{\Phi(L_1, M_1, T_1)}$$

and rearrange it. Doing so yields

$$\beta\Phi(L_1, M_1, T_1) = \Phi(L_2, M_2, T_2) \tag{2.50}$$

Dividing each term by its corresponding term in the first system of units, we get

$$\beta\Phi\left(\frac{L_1}{L_1}, \frac{M_1}{M_1}, \frac{T_1}{T_1}\right) = \Phi\left(\frac{L_2}{L_1}, \frac{M_2}{M_1}, \frac{T_2}{T_1}\right) \tag{2.51}$$

But

$$\Phi\left(\frac{L_1}{L_1}, \frac{M_1}{M_1}, \frac{T_1}{T_1}\right) = \Phi(1, 1, 1) = \kappa \tag{2.52}$$

where κ is a constant. Thus

$$\kappa\beta = \Phi\left(\frac{L_2}{L_1}, \frac{M_2}{M_1}, \frac{T_2}{T_1}\right) \tag{2.53}$$

Similarly for the third system of units

$$\kappa\gamma = \Phi\left(\frac{L_3}{L_1}, \frac{M_3}{M_1}, \frac{T_3}{T_1}\right) \tag{2.54}$$

Dividing conversion (2.54) by conversion (2.53) gives

$$\frac{\kappa\gamma}{\kappa\beta} = \frac{\gamma}{\beta} = \frac{\Phi(L_3/L_1, M_3/M_1, T_3/T_1)}{\Phi(L_2/L_1, M_2/M_1, T_2/T_1)} \tag{2.55}$$

Multiplying each term by its corresponding ratio of first system of units to second system of units gives

$$\frac{\gamma}{\beta}=\frac{\Phi\left(\frac{L_3}{L_1}\frac{L_1}{L_2},\frac{M_3}{M_1}\frac{M_1}{M_2},\frac{T_3}{T_1}\frac{T_1}{T_2}\right)}{\Phi\left(\frac{L_2}{L_1}\frac{L_1}{L_2},\frac{M_2}{M_1}\frac{M_1}{M_2},\frac{T_2}{T_1}\frac{T_1}{T_2}\right)}=\frac{\Phi\left(\frac{L_3}{L_1}\frac{L_1}{L_2},\frac{M_3}{M_1}\frac{M_1}{M_2},\frac{T_3}{T_1}\frac{T_1}{T_2}\right)}{\kappa}$$
$$=\frac{1}{\kappa}\Phi\left(\frac{L_3}{L_1}\frac{L_1}{L_2},\frac{M_3}{M_1}\frac{M_1}{M_2},\frac{T_3}{T_1}\frac{T_1}{T_2}\right) \tag{2.56}$$

and defining $\kappa = 1$, yields

$$\frac{\gamma}{\beta}=\Phi\left(\frac{L_3}{L_2},\frac{M_3}{M_2},\frac{T_3}{T_2}\right) \tag{2.57}$$

Equating Eq. (2.49) to Eq. (2.57) gives us

$$\frac{\Phi(L_3, M_3, T_3)}{\Phi(L_2, M_2, T_2)}=\Phi\left(\frac{L_3}{L_2},\frac{M_3}{M_2},\frac{T_3}{T_2}\right) \tag{2.58}$$

Differentiating Eq. (2.58) with respect to L_3 gives

$$\frac{\partial\Phi(L_3, M_3, T_3)/\partial L_3}{\Phi(L_2, M_2, T_2)}=\frac{1}{L_2}\Phi\left(\frac{L_3}{L_2},\frac{M_3}{M_2},\frac{T_3}{T_2}\right) \tag{2.59}$$

When we let $L_1 = L_2 = L_3$, $M_1 = M_2 = M_3$, and $T_1 = T_2 = T_3$, Eq. (2.59) becomes

$$\frac{d\Phi(L, M, T)/dL}{\Phi(L, M, T)}=\frac{1}{L}\Phi(1, 1, 1)=\frac{a}{L} \tag{2.60}$$

where $\Phi(111)$ is a constant designated as "a." Rearranging Eq. (2.60) gives

$$\frac{d\Phi(L, M, T)}{\Phi(L, M, T)}=a\left(\frac{dL}{L}\right) \tag{2.61}$$

Integrating yields

$$\ln(\Phi(L, M, T)) = a\ln(L) + \ln(\Phi'(M, T)) \tag{2.62}$$

or, in exponential notation

$$\Phi(L, M, T) = L^a\Phi'(M, T) \tag{2.63}$$

where $\Phi'(M,T)$ is a new function dependent upon M and T only. Performing the same operations on M and T eventually produces

$$\Phi(L, M, T) = L^a M^b T^c \tag{2.64}$$

Thus the dimension function which determines the physical magnitude is a monomial power law, as purported by Lord Rayleigh in 1877 [20–22].

DIMENSIONAL HOMOGENEITY

Consider a dependent variable y represented by a function of independent variables $x_1, x_2, x_3, \ldots, x_n$. This statement in mathematical notation is

$$y = f(x_1, x_2, x_3, \ldots, x_n) \tag{2.65}$$

Let us assume that the function is the sum of its independent variables, thus

$$y = x_1 + x_2 + x_3 + \cdots + x_n \tag{2.66}$$

If the function represents a physical equation, then each term in the function has a dimension associated with it, namely

$$y[LMT] = x_1[L_1M_1T_1] + x_2[L_2M_2T_2] + x_3[L_3M_3T_3] + \cdots + x_n[L_nM_nT_n] \tag{2.67}$$

Substituting for y yields

$$\begin{aligned}(x_1 + x_2 + x_3 + \cdots + x_n)[LMT] = x_1[L_1M_1T_1] + x_2[L_2M_2T_2] \\ + x_3[L_3M_3T_3] + \cdots + x_n[L_nM_nT_n]\end{aligned} \tag{2.68}$$

Expanding the terms to the left of the equality sign gives

$$x_1[LMT] + \cdots + x_n[LMT] = x_1[L_1M_1T_1] + \cdots + x_n[L_nM_nT_n] \tag{2.69}$$

Equating each term yields

$$\begin{aligned} x_1[LMT] &= x_1[L_1M_1T_1] \\ x_2[LMT] &= x_2[L_2M_2T_2] \\ \vdots \quad & \quad \vdots \\ x_n[LMT] &= x_n[L_nM_nT_n] \end{aligned} \tag{2.70}$$

But, from above

$$\alpha[\Psi] = \alpha[\mathrm{LMT}] = \Phi(\mathrm{L}, \mathrm{M}, \mathrm{T}) = \mathrm{L^a M^b T^c} \tag{2.71}$$

Thus, Eq. (2.70) become

$$\begin{aligned} \mathrm{L^a M^b T^c} &= \mathrm{L^{a_1} M^{b_1} T^{c_1}} \\ \mathrm{L^a M^b T^c} &= \mathrm{L^{a_2} M^{b_2} T^{c_2}} \\ \vdots &= \vdots \\ \mathrm{L^a M^b T^c} &= \mathrm{L^{a_n} M^{b_n} T^{c_n}} \end{aligned} \tag{2.72}$$

Equating like dimensions gives

$$\begin{aligned} &\mathrm{L^a} = \mathrm{L^{a_1}} = \cdots = \mathrm{L^{a_n}} \\ &\mathrm{M^b} = \mathrm{M^{b_1}} = \cdots = \mathrm{M^{b_n}} \\ &\mathrm{T} = \mathrm{T^{c_1}} = \mathrm{T^{c_n}} \end{aligned} \tag{2.73}$$

which shows that, when adding or subtracting, the dimensions L, M, and T on each term must be the same. In other words we can only add apples to apples or oranges to oranges ... we cannot add apples and oranges to get "orpels."

MATRIX FORMULATION OF DIMENSIONAL ANALYSIS

Consider a dependent variable y represented by a function of independent variables $x_1^{\mathrm{k}_1}, x_2^{\mathrm{k}_2}, \ldots, x_n^{\mathrm{k}_n}$, where the k's are constants. Mathematically

$$y = f(x_1^{\mathrm{k}_1}, x_2^{\mathrm{k}_2}, \ldots, x_n^{\mathrm{k}_n}) \tag{2.74}$$

If we assume the function is the multiplicative product of the independent variables, then

$$y = x_1^{\mathrm{k}_1} x_2^{\mathrm{k}_2} \cdots x_n^{\mathrm{k}_n} \tag{2.75}$$

If the function represents a physical equation, then

$$y[\mathrm{LMT}] = x_1^{\mathrm{k}_1}[\mathrm{L_1 M_1 T_1}]^{\mathrm{k}_1} x_2^{\mathrm{k}_2}[\mathrm{L_2 M_2 T_2}]^{\mathrm{k}_2} \cdots x_n^{\mathrm{k}_n}[\mathrm{L}_n \mathrm{M}_n \mathrm{T}_n]^{\mathrm{k}_n} \tag{2.76}$$

Substituting for y in Eq. (2.76) gives

$$(x_1^{\mathrm{k}_1} x_2^{\mathrm{k}_2} \cdots x_n^{\mathrm{k}_n})[\mathrm{LMT}] = x_1^{\mathrm{k}_1}[\mathrm{L_1 M_1 T_1}]^{\mathrm{k}_1} x_2^{\mathrm{k}_2}[\mathrm{L_2 M_2 T_2}]^{\mathrm{k}_2} \cdots x_n^{\mathrm{k}_n}[\mathrm{L}_n \mathrm{M}_n \mathrm{T}_n]^{\mathrm{k}_n} \tag{2.77}$$

Then dividing by $(x_1^{k_1} x_2^{k_2} \cdots x_n^{k_n})$ yields

$$\begin{aligned}[\mathrm{LMT}] &= \frac{x_1^{k_1}[\mathrm{L_1M_1T_1}]^{k_1} x_2^{k_2}[\mathrm{L_2M_2T_2}]^{k_2} \cdots x_n^{k_n}[\mathrm{L}_n\mathrm{M}_n\mathrm{T}_n]^{k_n}}{(x_1^{k_1} x_2^{k_2} \cdots x_n^{k_n})} \\ &= [\mathrm{L_1M_1T_1}]^{k_1}[\mathrm{L_2M_2T_2}]^{k_2} \cdots [\mathrm{L}_n\mathrm{M}_n\mathrm{T}_n]^{k_n}\end{aligned} \tag{2.78}$$

But, as previously stated

$$\alpha[\Psi] = \alpha[\mathrm{LMT}] = \Phi(\mathrm{L, M, T}) = \mathrm{L^a M^b T^c}$$

and

$$\alpha[\mathrm{LMT}]^{k_n} = \Phi(\mathrm{L, M, T})^{k_n} = (\mathrm{L^a M^b T^c})^{k_n} \tag{2.79}$$

Substituting Eq. (2.79) into Eq. (2.77) yields

$$\mathrm{L^a M^b T^c} = (\mathrm{L^{a_1} M^{b_1} T^{c_1}})^{k_1}(\mathrm{L^{a_2} M^{b_2} T^{c_2}})^{k_2} \cdots (\mathrm{L^{a_n} M^{b_n} T^{c_n}})^{k_n} \tag{2.80}$$

Equating the exponential terms for L, M, and T, respectively, gives

$$\begin{aligned} a &= a_1k_1 + a_2k_2 + \cdots + a_nk_n \\ b &= b_1k_1 + b_2k_2 + \cdots + b_nk_n \\ c &= c_1k_1 + c_2k_2 + \cdots + c_nk_n \end{aligned} \tag{2.81}$$

Note that in Eq. (2.81), we have n terms but only three equations. Therefore to solve this system of linear equations, we need to assume or assign values to $n-3$ terms. For convenience, let $n = 5$ in the above system of linear equations, then we have five unknowns and three equations; thus we need to assume values for two unknowns. Let us assume we know k_3, k_4, and k_5. We will assume values for k_1 and k_2. We represent k_1 and k_2 as

$$\begin{aligned} k_1 &= k_1 + 0 + 0 + \cdots + 0 \\ k_2 &= 0 + k_2 + 0 + \cdots + 0 \end{aligned} \tag{2.82}$$

Adding the k_1 and k_2 (2.82) to the original set of linear equation (2.81) gives us

$$\begin{aligned} k_1 &= k_1 + 0 + 0 + \cdots + 0 \\ k_2 &= 0 + k_2 + 0 + \cdots + 0 \\ a &= a_1k_1 + a_2k_2 + \cdots + a_nk_n \\ b &= b_1k_1 + b_2k_2 + \cdots + b_nk_n \\ c &= c_1k_1 + c_2k_2 + \cdots + c_nk_n \end{aligned} \tag{2.83}$$

which in matrix notation becomes

$$\begin{bmatrix} 1 & 0 & 0 & 0 & 0 \\ 0 & 1 & 0 & 0 & 0 \\ a_1 & a_2 & a_3 & a_4 & a_5 \\ b_1 & b_2 & b_3 & b_4 & b_5 \\ c_1 & c_2 & c_3 & c_4 & c_5 \end{bmatrix} \begin{bmatrix} k_1 \\ k_2 \\ k_3 \\ k_4 \\ k_5 \end{bmatrix} = \begin{bmatrix} k_1 \\ k_2 \\ a \\ b \\ c \end{bmatrix} \tag{2.84}$$

From matrix algebra, we can partition the above matrices into the Identity matrix

$$I = \begin{bmatrix} 1 & 0 \\ 0 & 1 \end{bmatrix} \tag{2.85}$$

and the Zero matrix [23]

$$0 = \begin{bmatrix} 0 & 0 & 0 \\ 0 & 0 & 0 \end{bmatrix} \tag{2.86}$$

The matrix

$$\begin{bmatrix} a_1 & a_2 & a_3 & a_4 & a_5 \\ b_1 & b_2 & b_3 & b_4 & b_5 \\ c_1 & c_2 & c_3 & c_4 & c_5 \end{bmatrix} \tag{2.87}$$

is the Dimension matrix. It follows directly from the Dimension Table. The Dimension Table catalogs the dimensions of each variable of the original function. Thus, the Dimension Table has the below format

Variable		x_1	x_2	x_3	x_4	x_5
Dimension	L	a_1	a_2	a_3	a_4	a_5
	M	b_1	b_2	b_3	b_4	b_5
	T	c_1	c_2	c_3	c_4	c_5

The Dimension matrix can be partitioned into two matrices, one being a square matrix, that is, a matrix with the same number of rows as columns; the other being the bulk, or remaining matrix elements. We define the square matrix as the Rank matrix and the remaining matrix as the Bulk matrix. Partitioning the above Dimension matrix gives

$$\left[\begin{bmatrix} a_1 & a_2 \\ b_1 & b_2 \\ c_1 & c_2 \end{bmatrix} \begin{bmatrix} a_3 & a_4 & a_5 \\ b_3 & b_4 & b_5 \\ c_3 & c_4 & c_5 \end{bmatrix} \right] \tag{2.88}$$

where the Bulk matrix is

$$B = \begin{bmatrix} a_1 & a_2 \\ b_1 & b_2 \\ c_1 & c_2 \end{bmatrix} \tag{2.89}$$

and the Rank matrix is

$$R = \begin{bmatrix} a_3 & a_4 & a_5 \\ b_3 & b_4 & b_5 \\ c_3 & c_4 & c_5 \end{bmatrix} \tag{2.90}$$

We use the Rank matrix to calculate the "rank" of the Dimension matrix. We need the rank of the Dimension matrix to determine the number of independent solutions that exist for our system of linear equations. From linear algebra, the rank of a matrix is the number of linearly independent rows, or columns, of a matrix [24]. In other words the rank of a matrix is the number of independent equations in a system of linear equations. Thus the number of variables in a system of linear equations, that is, the number of columns in the Dimension matrix minus the rank of the Dimension matrix equals the number of selectable unknowns [23]. Mathematically

$$N_{\text{Var}} - R = N_{\text{Specified}} \tag{2.91}$$

where N_{Var} is the number of variables in our analysis, R is the rank of the Dimension matrix, and $N_{\text{Specified}}$ is the number of variables we need to specify in our analysis.

To determine the Rank of the Dimension matrix, we must calculate the determinant of the Rank matrix. If the determinant of the Rank matrix is nonzero, then R is the number of rows or the number of columns in the Rank Matrix. The above Rank matrix is a 3×3 matrix; therefore the rank of its Dimension matrix is 3. In this case, $N_{\text{Var}} = 5$ and $R = 3$; therefore $N_{\text{Var}} - R = N_{\text{Specified}}$ is $5 - 3 = 2$. Therefore, to solve the above set of linear equations, we need to specify two unknowns.

We can now rewrite the matrix equation as

$$\begin{bmatrix} 1 & 0 & 0 & 0 & 0 \\ 0 & 1 & 0 & 0 & 0 \\ a_1 & a_2 & a_3 & a_4 & a_5 \\ b_1 & b_2 & b_3 & b_4 & b_5 \\ c_1 & c_2 & c_3 & c_4 & c_5 \end{bmatrix} \begin{bmatrix} k_1 \\ k_2 \\ k_3 \\ k_4 \\ k_5 \end{bmatrix} = \begin{bmatrix} k_1 \\ k_2 \\ a \\ b \\ c \end{bmatrix}$$

in terms of the partitioned matrices; the above matrix equation becomes

$$\begin{bmatrix} I & 0 \\ B & R \end{bmatrix} \begin{bmatrix} k_1 \\ k_2 \\ k_3 \\ k_4 \\ k_5 \end{bmatrix} = \begin{bmatrix} k_1 \\ k_2 \\ a \\ b \\ c \end{bmatrix} \tag{2.92}$$

where B is

$$\begin{bmatrix} a_1 & a_2 \\ b_1 & b_2 \\ c_1 & c_2 \end{bmatrix} \tag{2.93}$$

and R is

$$\begin{bmatrix} a_3 & a_4 & a_5 \\ b_3 & b_4 & b_5 \\ c_3 & c_4 & c_5 \end{bmatrix} \tag{2.94}$$

The solution to the matrix equation is [23]

$$\begin{bmatrix} k_1 \\ k_2 \\ k_3 \\ k_4 \\ k_5 \end{bmatrix} = \begin{bmatrix} I & 0 \\ -R^{-1}B & R^{-1} \end{bmatrix} \begin{bmatrix} k_1 \\ k_2 \\ a \\ b \\ c \end{bmatrix} \tag{2.95}$$

where

$$T = \begin{bmatrix} I & 0 \\ -R^{-1}B & R^{-1} \end{bmatrix} \tag{2.96}$$

is the Total matrix.

With regard to dimensional analysis, the number of columns in the Dimension matrix equals the number of variables in the system of linear equations and the difference between the number of columns in the Dimension matrix and the rank of the Dimension matrix equals the number of selectable unknowns in the system of linear equations. The number of selectable unknowns equals the number of columns in the Identity matrix. The product of reading down a column of the Identity matrix is a dimensionless parameter. The number of dimensionless parameters generated by a given chemical process is

$$N_P = N_{Var} - R \tag{2.97}$$

where N_P is the number of independent dimensional or dimensionless parameters obtainable from a given set of linear equations. This result is known as Buckingham's Theorem or the Pi Theorem [20].

IDENTIFYING VARIABLES FOR DIMENSIONAL ANALYSIS

The question arises: how do we identify the variables for a Dimensional Analysis study? The best way to identify the variables for use in a Dimensional Analysis is to write the conservation laws and constitutive equations underpinning the process being studied. Constitutive equations describe a specific response of a given variable to an external force. The most familiar constitutive equations are Newton's law of viscosity, Fourier's law of heat conduction, and Fick's law of diffusion.

The issue when identifying variables for a Dimensional Analysis is not having too many, but missing pertinent ones. In the former situation we still obtain the correct result; however, that result will contain extraneous variables, variables not actually required by Dimensional Analysis. In the latter situation Dimensional Analysis produces an incorrect result. Therefore, to ensure the correct result, we will include any variable we deem remotely pertinent to the process being investigated. We can then identify, during our analysis of the process, which variables are irrelevant.

SUMMARY

This chapter provided the foundation for using a matrix-based procedure for identifying and defining dimensionless parameters.

REFERENCES

[1] R. Pankhurst, Dimensional Analysis and Scale Factors, Chapman and Hall, Ltd, London, UK, 1964, p. 13.

[2] R. Pankhurst, Dimensional Analysis and Scale Factors, Chapman and Hall, Ltd, London, UK, 1964, p. 20.

[3] H. Hornung, Dimensional Analysis: Examples of the Use of Symmetry, Dover Publications, Inc, Mineola, NY, 2006, p. 1.

[4] D. Ipsen, Units, Dimensions, and Dimensionless Numbers, McGraw-Hill Book Company, Inc, New York, NY, 1960 (Chapter 2).

[5] R. Pankhurst, Dimensional Analysis and Scale Factors, Chapman and Hall, Ltd, London, UK, 1964, pp. 40–43.

[6] R. Tolman, The principle of similtude and the principle of dimensional homogeneity, Phys. Rev. 9 (1917) 237.

[7] G. Lewis, M. Randall, Thermodynamics and Free Energy of Chemical Substances, McGraw-Hill Book Company, Inc, New York, NY, 1923, p. 3.

[8] K. Denbigh, The Principles of Chemical Equilibrium, third ed., Cambridge University Press, Cambridge, UK, 1978, p. 7.

[9] A. Klinkenberg, The American engineering system of units and its dimensional constant g_c, Ind. Eng. Chem. 61 (4) (1969) 53.

[10] R. Pankhurst, Dimensional Analysis and Scale Factors, Chapman and Hall, Ltd, London, UK, 1964, p. 25.

[11] A. Klinkenberg, Dimensional systems and systems of units in physics with special reference to chemical engineering, Chem. Eng. Sci. 4 (1955) 180.

[12] Not to be confused with the language "Old English". In our case, Old English Engineering means units devised during the 19th Century and used well into the mid-20th Century.

[13] G. Murphy, Similitude in Engineering, The Ronald Press Company, New York, NY, 1950, p. 7.

[14] P. Bridgman, Dimensional Analysis, Yale University Press, New Haven, CT, 1922, p. 42.

[15] J. Hunsaker, B. Rightmere, Engineering Applications of Fluid Mechanics, McGraw-Hill Book Company, New York, NY, 1947 (Chapter 7).

[16] H. Huntley, Dimensional Analysis, Dover Publications, Inc, New York, NY, 1967 (Chapter 1; originally published by McDonald and Company, Ltd, 1952).

[17] G. Barenblatt, Scaling, Cambridge University Press, Cambridge, UK, 2003, pp. 17–20.

[18] G. Barenblatt, Scaling, Self-similarity, and Intermediate Asymptotics, Cambridge University Press, Cambridge, UK, 1996, pp. 34–37.

[19] H. Langhaar, Dimensional Analysis and Theory of Models, John Wiley and Sons, Inc, New York, NY, 1951 (Chapter 4)

[20] E. Buckingham, On physically similar systems, Phys. Rev. 4 (1914) 354.

[21] L. Rayleigh, Letter to the Editor, Nature 95 (1915) 66.

[22] R. Tolman, The principle of similitude, Phys. Rev. 3 (1914) 244.

[23] T. Szirtes, Applied Dimensional Analysis and Modeling, second ed., Butterworth-Heinemann, Burlingham, MA, 2007 (Chapter 7).

[24] M. Jain, Vector Spaces and Matrices in Physics, Narosa Publishing House, New Delhi, India, 2001, p. 75.

CHAPTER 3

Control Regime

INTRODUCTION

Every chemical process consists of a number of unit operations, which includes

- feed storage;
- feed purification;
- reaction;
- product separation and purification;
- product storage.

Feed purification generally involves absorption, adsorption, extraction, and/or distillation. Reaction involves agitated batch, agitated semibatch, continuous stirred tank, or continuous flow reactors. The continuous flow reactors may be empty or contain a mass of solid catalyst. Product separation and purification involves distillation in the petrochemical industry or extraction and crystallization in the extractive metallurgy and pharmaceutical industries; absorption is used to a lesser extent.

We can envision a mechanism of one or more steps for each of these unit operations and we can write a rate equation for each step. We can then relate each of these individual rate equations to an overall rate constant. For a mechanism with two or more steps in series, one step will be slower than the other steps: we say this slow step is the rate controlling step. For example, a gas–liquid reaction in a laboratory-sized reactor is either heat transfer controlled or reaction rate controlled. If we cannot supply heat fast enough to maintain the reaction or if we cannot remove heat fast enough to control the reaction, we say the reaction is heat transfer controlled. If, on the other hand, we can supply or remove heat faster than required by the reaction, then we say the reaction is reaction rate controlled. In general, laboratory-sized batch and semibatch reactors have large heat transfer surface area to reaction volume ratios; therefore, transferring heat to

Scaling Chemical Processes.
© 2016 Elsevier Inc. All rights reserved.

or from the reactor is not an issue. In other words, most laboratory-sized batch and semibatch reactors are not heat transfer rate controlled; rather, they are reaction rate controlled. For commercial-sized batch and semibatch reactors, the opposite is true: they are generally heat transfer rate controlled because reaction volume grows faster than does heat transfer surface area; thus transferring heat to or from such a reactor is the rate controlling mechanism or step. Pilot plant-sized batch and semibatch reactors fall between these two extremes, depending upon how much heat transfer surface area the reactor possesses.

CONTROL REGIMES FOR HEAT TRANSFER

Consider a jacketed vessel containing an agitated liquid. We heat the liquid inside this vessel by condensing steam in the jacket. The condensed steam forms a liquid film on the wall separating the jacket from the vessel. Heat flows through this layer of condensed steam, then through the metal wall. Note that the outside wall of the vessel is covered with a layer of deposit and scale; the inside wall of the vessel is also covered by a thin layer of scale. Both layers impede the flow of heat into the vessel. When a fluid flows over an interface, such as a metal or gas–liquid or liquid–liquid surface, a boundary layer of that fluid forms on the surface. Such is the case with the agitated liquid in this vessel. This liquid film will be "thick" if the liquid has a high viscosity and we agitate it at a "low" rate, and this liquid film will be "thin" if the liquid has a low viscosity and we agitate it at a "high" rate. Thus convection, that is, fluid movement, controls liquid film thickness on the inside wall, which, in turn, impedes the rate of heat passing through it and into the bulk liquid in our vessel.

The rate at which heat moves from the condensing steam to the condensate film is

$$Q = h_{\mathrm{Steam}} A_{\mathrm{Scale}} (T_{\mathrm{Steam}} - T_{\mathrm{Cond}}) \tag{3.1}$$

where Q is the rate of heat flow $[\mathrm{L^2MT^{-3}}]$; h_{Steam} is the heat transfer film coefficient for condensed steam $[\mathrm{MT^{-3}\theta^{-1}}]$; A_{Scale} is the area of the scale covering the vessel's outside wall $[\mathrm{L^2}]$; T_{Steam} is the temperature of the steam $[\theta]$; and T_{Cond} is the temperature of the condensate covering the scale on the vessel wall $[\theta]$.

The rate at which heat moves through the scale layer is

$$Q = h_{\text{Scale,Out}} A_{\text{Wall,Out}} (T_{\text{Cond}} - T_{\text{Wall,Out}}) \tag{3.2}$$

where $h_{\text{Scale,Out}}$ is the heat transfer film coefficient for the scale layer [MT$^{-3}\theta^{-1}$]; $A_{\text{Wall,Out}}$ is the area of the outside wall of the vessel [L^2]; and $T_{\text{Wall,Out}}$ is the temperature of the outside wall of the vessel [θ]. All other variables have been defined earlier.

The rate of heat flow through the vessel's metal wall is

$$Q = \left(\frac{k}{x}\right) A_{\text{Wall,Ave}} (T_{\text{Wall,Out}} - T_{\text{Wall,In}}) \tag{3.3}$$

where k is the thermal conductivity of the metal wall [LMT$^{-3}\theta^{-1}$]; x is the wall thickness of the vessel [L]; $A_{\text{Wall,Ave}}$ is the average area of the wall of the vessel [L^2]; and $T_{\text{Wall,In}}$ is the temperature of the inside wall of the vessel [θ]. All other variables are defined earlier.

The flow of heat through the scale covering the inside wall of the vessel is

$$Q = h_{\text{Scale,In}} A_{\text{Wall,In}} (T_{\text{Wall,In}} - T_{\text{Scale,In}}) \tag{3.4}$$

where $h_{\text{Scale,In}}$ is the heat transfer film coefficient for the inside scale layer [MT$^{-3}\theta^{-1}$]; $A_{\text{Wall,In}}$ is the area of the inside wall of the vessel [L^2]; and $T_{\text{Scale,In}}$ is the temperature of the inside wall of the vessel [θ]. All other variables have been defined earlier.

The rate of heat flow into the agitated liquid is

$$Q = h_{\text{Liq}} A_{\text{Liq}} (T_{\text{Scale,In}} - T_{\text{Bulk}}) \tag{3.5}$$

where h_{Liq} is the heat transfer film coefficient for the boundary layer inside the vessel [MT$^{-3}\theta^{-1}$]; A_{Liq} is the area of the boundary layer inside the vessel [L^2]; and T_{Bulk} is the temperature of the agitated, bulk liquid inside the vessel [θ]. All other variables are defined earlier.

Unfortunately, of all the above-mentioned temperatures, we only know T_{Steam} and T_{Bulk} with any accuracy. Therefore, we must derive a heat transfer rate equation in terms of T_{Steam} and T_{Bulk}. We achieve that goal in the following way. Solving Eq. (3.2) for T_{Cond}, then substituting the result into Eq. (3.1) gives us

$$\frac{Q}{h_{\text{Steam}} A_{\text{Scale}}} + \frac{Q}{h_{\text{Scale,Out}} A_{\text{Wall,Out}}} + T_{\text{Wall,Out}} = T_{\text{Steam}} \tag{3.6}$$

Solving Eq. (3.3) for $T_{\text{Wall,Out}}$, then substituting into Eq. (3.6) yields

$$\frac{Q}{h_{\text{Steam}}A_{\text{Scale}}}+\frac{Q}{h_{\text{Scale,Out}}A_{\text{Wall,Out}}}+\frac{Q}{(k/x)A_{\text{Wall,Ave}}}+T_{\text{Wall,In}}=T_{\text{Steam}} \tag{3.7}$$

Substituting for $T_{\text{Wall,In}}$ in Eq. (3.7) provides us with

$$\frac{Q}{h_{\text{Steam}}A_{\text{Scale}}}+\frac{Q}{h_{\text{Scale,Out}}A_{\text{Wall,Out}}}+\frac{Q}{(k/x)A_{\text{Wall,Ave}}}+\frac{Q}{h_{\text{Scale,In}}A_{\text{Wall,In}}}+T_{\text{Scale,In}}=T_{\text{Steam}} \tag{3.8}$$

And, finally, solving for $T_{\text{Scale,In}}$, then substituting the result into Eq. (3.8) gives us

$$\frac{Q}{h_{\text{Steam}}A_{\text{Scale}}}+\frac{Q}{h_{\text{Scale,Out}}A_{\text{Wall,Out}}}+\frac{Q}{(k/x)A_{\text{Wall,Ave}}}+\frac{Q}{h_{\text{Scale,In}}A_{\text{Wall,In}}}+\frac{Q}{h_{\text{Liq}}A_{\text{Liq}}}=T_{\text{Steam}}-T_{\text{Bulk}} \tag{3.9}$$

Note that all the Qs are equal because no rate can be greater than the slowest rate in the series. Thus

$$\left\{\frac{1}{h_{\text{Steam}}A_{\text{Scale}}}+\frac{1}{h_{\text{Scale,Out}}A_{\text{Wall,Out}}}+\frac{1}{(k/x)A_{\text{Wall,Ave}}}+\frac{1}{h_{\text{Scale,In}}A_{\text{Wall,In}}}+\frac{1}{h_{\text{Liq}}A_{\text{Liq}}}\right\}Q=T_{\text{Steam}}-T_{\text{Bulk}} \tag{3.10}$$

The rate of heat flow for the vessel as a whole is

$$Q=UA(T_{\text{Steam}}-T_{\text{Bulk}}) \tag{3.11}$$

where U is an overall heat transfer rate coefficient $[\text{MT}^{-3}\theta^{-1}]$ and A is a defined reference area $[\text{L}^2]$. If we designate A as the inside liquid contact area, then Eq. (3.11) becomes

$$Q=U_{\text{Liq,In}}A_{\text{Liq,In}}(T_{\text{Steam}}-T_{\text{Bulk}}) \tag{3.12}$$

Rearranging Eq. (3.12)

$$\frac{Q}{U_{\text{Liq,In}}A_{\text{Liq,In}}} = (T_{\text{Steam}} - T_{\text{Bulk}}) \tag{3.13}$$

then comparing the result with Eq. (3.10) shows that

$$\frac{Q}{U_{\text{Liq,In}}A_{\text{Liq,In}}} = \left\{ \frac{1}{h_{\text{Steam}}A_{\text{Scale}}} + \frac{1}{h_{\text{Scale,Out}}A_{\text{Wall,Out}}} + \frac{1}{(k/x)A_{\text{Wall,Ave}}} + \frac{1}{h_{\text{Scale,In}}A_{\text{Wall,In}}} + \frac{1}{h_{\text{Liq}}A_{\text{Liq}}} \right\} Q \tag{3.14}$$

or

$$\frac{1}{U_{\text{Liq,In}}} = \left\{ \frac{1}{h_{\text{Steam}}A_{\text{Scale}}} + \frac{1}{h_{\text{Scale,Out}}A_{\text{Wall,Out}}} + \frac{1}{(k/x)A_{\text{Wall,Ave}}} + \frac{1}{h_{\text{Scale,In}}A_{\text{Wall,In}}} + \frac{1}{h_{\text{Liq}}A_{\text{Liq}}} \right\} A_{\text{Liq,In}} \tag{3.15}$$

In most cases, we assume all the areas are nearly equal; therefore

$$\frac{1}{U_{\text{Liq,In}}} = \frac{1}{h_{\text{Steam}}} + \frac{1}{h_{\text{Scale,Out}}} + \frac{1}{(k/x)} + \frac{1}{h_{\text{Scale,In}}} + \frac{1}{h_{\text{Liq}}} \tag{3.16}$$

which expresses a series of resistances as a sum. One of these resistances will regulate the rate of heat flow to or from an agitated vessel. If

$$h_{\text{Steam}} > k/x > h_{\text{Liq}} > h_{\text{Scale,In}} \gg h_{\text{Scale,Out}} \tag{3.17}$$

then the resistance $1/h_{\text{Scale,Out}}$ controls the rate of heat flow to or from an agitated vessel. In such a situation, scaling the vessel up or down with respect to heat transfer must be done using $h_{\text{Scale,Out}}$ as the primary heat transfer film coefficient. If h_{Liq} describes the rate controlling step, then the scaling effort must be done with respect to h_{Liq}. Unfortunately, many scaling efforts are done without the knowledge of the rate controlling step of the unit operation. When this situation occurs, the prototype will have a different operating behavior than the model.

CONTROL REGIMES FOR GAS–LIQUID SEMIBATCH PROCESSES

When we react a gas with a liquid or react two immiscible liquids, mass transfer occurs simultaneously with reaction. In such a situation, chemical kinetics is coupled with mass transport, which means we are

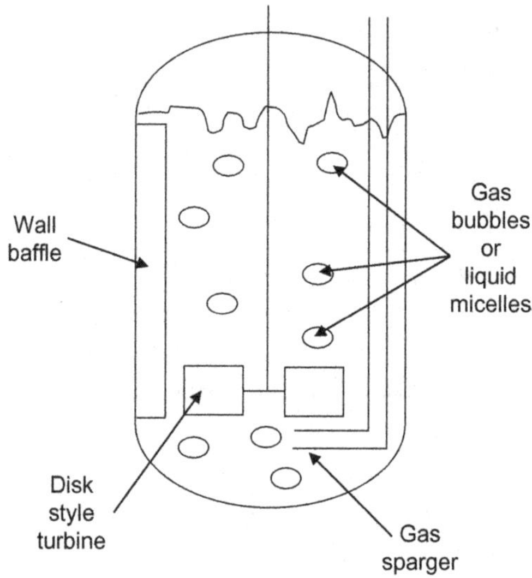

Figure 3.1 Macroscopic features of a gas–liquid semibatch process.

no longer conducting a reaction, rather, we are managing a complex chemical–physical process. Such processes are usually diffusion rate controlled, especially if the gas is sparingly soluble in the liquid or if one liquid is sparingly soluble in a second liquid. Therefore, before scaling such a process, we need to determine whether the process is mass transfer rate controlled or reaction rate controlled.

Fig. 3.1 presents the macroscopic features of a gas–liquid semibatch process. We completely charge the liquid reactant, either pure or diluted in an inert solvent, into a reactor equipped with a gas sparger located near or under the agitator. The agitator can be a propeller or one of the many turbine designs available to us. We generally install four wall baffles at 90 degree separation in the reactor to suppress vortex formation in the liquid during agitation. We use the agitation and baffles to evenly distribute the gas reactant throughout the liquid reactant.

Fig. 3.2 shows the microscopic features of a gas–liquid process. Each gas bubble contains gaseous reactant. If the bulk fluid vaporizes into the gas bubbles or if the gas is diluted with an inert gas, then a stagnant film of gas molecules develops adjacent to the gas–liquid interface even though some circular motion of gas molecules occurs inside the gas bubble [1]. A stagnant film inside the gas bubble does

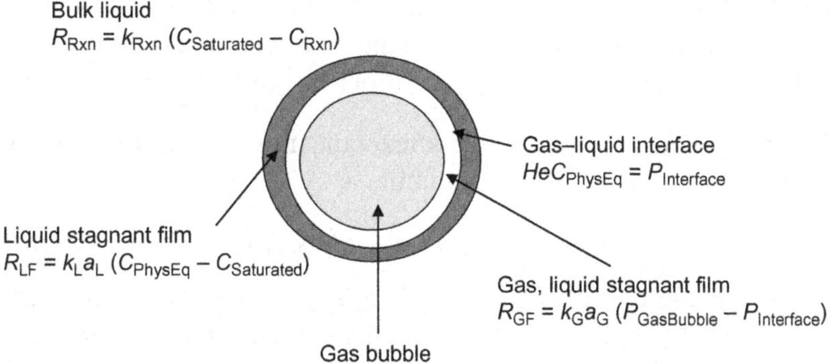

Figure 3.2 Microscopic features of gas–liquid or immiscible liquids semibatch process.

not form if the bulk fluid does not volatilize into the gas bubbles or if the gas is pure. For this analysis, we assume a stagnant film forms inside the gas bubble [2]. Diffusion governs the movement of gas molecules within this stagnant gas film. Physical equilibrium, expressed as Henry's law, controls the movement of gas reactant from the gas phase to the liquid phase. A stagnant liquid film surrounds each gas–liquid interface and diffusion governs the movement of solubilized gas within it. The bulk phase of the liquid reactant comprises the fluid moving between each gas bubble. The agitator provides the energy necessary to induce liquid motion within the reactor.

We describe, mathematically, the rate at which the gas reactant moves across the stagnant gas film as

$$R_{GF} = k_G a_G (P_{GasBubble} - P_{Interface}) \tag{3.18}$$

where R_{GF} is the rate of gas molecules moving across the stagnant gas film inside each gas bubble (kg/m · s^3). k_G is the velocity at which gas molecules cross the stagnant gas film (m/s); a_G is the surface area to volume of the stagnant gas film adjacent to the gas–liquid interface (m^2/m^3); and $P_{GasBubble}$ and $P_{Interface}$ is the pressure of the gas bubble and the pressure of the gas reactant at the gas–liquid interface, respectively (kg/m · s^2).

Henry's law controls the amount of gas dissolved by the liquid reactant; it is

$$HeC_{PhysEq} = P_{Interface} \tag{3.19}$$

where He is Henry's constant ($kg \cdot m^2/moles \cdot s^2$) and C_{PhysEq} is the concentration of dissolved gas reactant in liquid reactant at physical equilibrium between the two phases ($moles/m^3$).

The rate at which dissolved gas reactant moves across the stagnant liquid film surrounding each gas bubble is

$$R_{LF} = k_L a_L (C_{PhysEq} - C_{Saturated}) \tag{3.20}$$

where R_{LF} is the rate of dissolved gas moving across the stagnant film ($moles/m^3 \cdot s$); k_L is the velocity at which dissolved gas moves across the stagnant liquid film (m/s); a_L is the surface area to volume of the stagnant liquid film adjacent to the gas–liquid interface (m^2/m^3); and $C_{Saturated}$ is the concentration of dissolved gas in bulk liquid ($moles/m^3$).

The rate of chemical reaction within the bulk liquid is

$$R_{Rxn} = k_{Rxn}(C_{Saturated} - C_{Rxn}) \tag{3.21}$$

where R_{Rxn} is the rate of chemical reaction ($moles/m^3 \cdot s$); k_{Rxn} is the kinetic rate constant (s^{-1}); and C_{Rxn} is the concentration of dissolved gas reactant at process conditions ($moles/m^3$).

Unfortunately, in Eqs. (3.18) through (3.21), we only know $P_{GasBubble}$ and C_{Rxn}. $P_{GasBubble}$ is the pressure of the gas as it enters the semibatch reactor and C_{Rxn} is the dissolved gas in the liquid at process conditions, which we measure by removing a sample from the reactor. Thus we must reduce the above set of rate equations to one equation containing $P_{GasBubble}$ and C_{Rxn}. Substituting Henry's law for $P_{Interface}$ gives us

$$R_{GF} = k_G a_G (P_{GasBubble} - He C_{PhysEq}) \tag{3.22}$$

which, upon rearranging, becomes

$$\frac{R_{GF}}{He k_G a_G} = \left(\frac{P_{GasBubble}}{He} - C_{PhysEq}\right) \tag{3.23}$$

Solving the R_{LF} rate equation for C_{PhysEq}, then substituting into Eq. (3.23) provides

$$\frac{R_{GF}}{He k_G a_G} = \left(\frac{P_{GasBubble}}{He} - \frac{R_{LF}}{k_L a_L} - C_{Saturated}\right) \tag{3.24}$$

Rearranging Eq. (3.24) gives

$$\frac{R_{GF}}{Hek_{G}a_{G}} + \frac{R_{LF}}{k_{L}a_{L}} = \left(\frac{P_{GasBubble}}{He} - C_{Saturated}\right) \tag{3.25}$$

Solving the R_{Rxn} rate equation for $C_{Saturated}$, then substituting into Eq. (3.25) yields

$$\frac{R_{GF}}{Hek_{G}a_{G}} + \frac{R_{LF}}{k_{L}a_{L}} = \left(\frac{P_{GasBubble}}{He} - \frac{R_{Rxn}}{k_{Rxn}} - C_{Rxn}\right) \tag{3.26}$$

and upon rearranging, we obtain

$$\frac{R_{GF}/He}{k_{G}a_{G}} + \frac{R_{LF}}{k_{L}a_{L}} + \frac{R_{Rxn}}{k_{Rxn}} = \left(\frac{P_{GasBubble}}{He} - C_{Rxn}\right) \tag{3.27}$$

We now have all the rates related to the two variables we can measure. We know the overall rate of the process is

$$R_{Overall} = k_{Overall}\left(\frac{P_{GasBubble}}{He} - C_{Rxn}\right) \tag{3.28}$$

where $R_{Overall}$ is the overall rate (moles/m$^{3}\cdot$s) for the semibatch process and $k_{Overall}$ is the overall rate constant for the semibatch process (s^{-1}).

Since none of the rates can be faster than the slowest rate and the overall rate equals the slowest rate, we can write

$$\frac{R_{G}}{He} = R_{LF} = R_{Rxn} = R_{Overall} \tag{3.29}$$

Therefore

$$\frac{R_{Overall}}{k_{G}a_{G}} + \frac{R_{Overall}}{k_{L}a_{L}} + \frac{R_{Overall}}{k_{Rxn}} = \left(\frac{P_{GasBubble}}{He} - C_{Rxn}\right) \tag{3.30}$$

or

$$R_{Overall}\left(\frac{1}{k_{G}a_{G}} + \frac{1}{k_{L}a_{L}} + \frac{1}{k_{Rxn}}\right) = \left(\frac{P_{GasBubble}}{He} - C_{Rxn}\right) \tag{3.31}$$

Rearranging Eq. (3.31) gives us

$$R_{Overall} = \frac{1}{\left(\frac{1}{k_{G}a_{G}} + \frac{1}{k_{L}a_{L}} + \frac{1}{k_{Rxn}}\right)}\left(\frac{P_{GasBubble}}{He} - C_{Rxn}\right) \tag{3.32}$$

which means

$$k_{\mathrm{Overall}} = \frac{1}{\left(\frac{1}{k_{\mathrm{G}}a_{\mathrm{G}}} + \frac{1}{k_{\mathrm{L}}a_{\mathrm{L}}} + \frac{1}{k_{\mathrm{Rxn}}}\right)} \tag{3.33}$$

or

$$\frac{1}{k_{\mathrm{Overall}}} = \frac{1}{k_{\mathrm{G}}a_{\mathrm{G}}} + \frac{1}{k_{\mathrm{L}}a_{\mathrm{L}}} + \frac{1}{k_{\mathrm{Rxn}}} \tag{3.34}$$

Note that in Eq. (3.34), k_{Overall}, k_{G}, and k_{L} are constants. k_{Rxn} denotes the reaction rate that occurs in unit time, which depends upon the pressure and temperature at which we operate the semibatch reactor. Since, for most reactions, we operate semibatch reactors at constant pressure and temperature, k_{Rxn} does not change. k_{G} and k_{L} depend upon the mean free path of a molecule through the gas phase and through the liquid phase, respectively. Both mean free paths depend upon the density of their respective phases. If the phase densities are constant, then k_{G} and k_{L} will be constant. Therefore, any fluctuation in k_{Overall} must arise from fluctuations in a_{G} and a_{L}.

If the gas bubbles are spherical, then

$$a_{\mathrm{G}} = \frac{4\pi r^2}{(4/3)\pi r^3} = \frac{3}{r} \tag{3.35}$$

Substituting this result into the resistance equation (3.34) yields

$$\frac{1}{k_{\mathrm{Overall}}} = \frac{r}{3k_{\mathrm{G}}} + \frac{1}{k_{\mathrm{L}}a_{\mathrm{L}}} + \frac{1}{k_{\mathrm{Rxn}}} \tag{3.36}$$

Therefore, as gas bubble radius decreases, k_{Overall} increases.

Injecting the smallest possible gas bubble size into a semibatch reactor ensures the greatest possible k_{Overall} with regard to the gas phase reactant. If we inject the smallest size gas bubble into the liquid phase and assume the average gas bubble size in the liquid phase is constant, then a_{G} is essentially constant. This assertion allows us to rearrange Eq. (3.34) for $1/k_{\mathrm{Overall}}$. Multiplying the first term left of the equal sign by 1; that is, by $k_{\mathrm{Rxn}}/k_{\mathrm{Rxn}}$, then multiplying the leftmost term by 1; or, $k_{\mathrm{G}}a_{\mathrm{G}}/k_{\mathrm{G}}a_{\mathrm{G}}$ yields

$$\frac{1}{k_{\mathrm{Overall}}} = \frac{k_{\mathrm{Rxn}}}{k_{\mathrm{G}}a_{\mathrm{G}}k_{\mathrm{Rxn}}} + \frac{1}{k_{\mathrm{L}}a_{\mathrm{L}}} + \frac{k_{\mathrm{G}}a_{\mathrm{G}}}{k_{\mathrm{Rxn}}k_{\mathrm{G}}a_{\mathrm{G}}} \tag{3.37}$$

Combining terms gives us

$$\frac{1}{k_{\text{Overall}}} = \frac{k_{\text{Rxn}} + k_{\text{G}}a_{\text{G}}}{k_{\text{G}}a_{\text{G}}k_{\text{Rxn}}} + \frac{1}{k_{\text{L}}a_{\text{L}}} \tag{3.38}$$

which we recognize as the equation for a straight line if we define the ordinate as $1/k_{\text{Overall}}$ and the abscissa as $1/a_{\text{L}}$ since $1/k_{\text{L}}$ is a constant. The intercept for this straight line is

$$(k_{\text{Rxn}} + k_{\text{G}}a_{\text{G}})/k_{\text{G}}a_{\text{G}}k_{\text{Rxn}}$$

a_{L} is the surface area to volume ratio of the stagnant liquid film surrounding each gas bubble in the liquid phase of the semibatch reactor. It is well established by fluid dynamics that the thickness of a stagnant gas or liquid film adjacent to a surface decreases as the velocity of the fluid flowing over it increases. Mathematically

$$\delta \propto \frac{1}{\sqrt{Re}} = \frac{1}{\sqrt{\rho D v/\mu}} = \sqrt{\frac{\mu}{\rho D v}} \tag{3.39}$$

where Re is the Reynolds number in the direction of fluid flow; ρ is the density of the flowing fluid (kg/m^3); D is the diameter of the conduit through which the fluid flows (m); v is the velocity of the flowing fluid (m/s); and μ is the viscosity of the flowing fluid (kg/m · s) [3]. If we postulate each gas bubble to be spherical, then a_{L} is

$$a_{\text{L}} = \frac{4\,\pi\,\delta^2}{(4/3)\pi\,\delta^3} = \frac{3}{\delta} \tag{3.40}$$

Thus k_{Overall} is a function of $k_{\text{L}}a_{\text{L}}$ and that $k_{\text{L}}a_{\text{L}}$ is a function of Reynolds number. In other words

$$\begin{aligned} k_{\text{Overall}} &= f(k_{\text{L}}a_{\text{L}}) \\ k_{\text{L}}a_{\text{L}} &= f(Re) \end{aligned} \tag{3.41}$$

A relationship between k_{Overall} and Re, represented as agitator RPMs, has been intuitively surmised since the inception of chemical engineering as a separate professional discipline. One semibatch variable available to the early chemical engineer was agitator speed or agitator RPMs. By increasing agitator RPM, pioneer chemical engineers found that k_{Overall} increased, then plateaued. Fig. 3.3 shows this common result. We generally say a semibatch process demonstrating a linear response between k_{Overall} and agitator RPM is "diffusion" rate controlled; we say a plateaued response represents a "reaction" rate controlled semibatch process.

If we assume that $a_L \propto$ RPM, then we can write Eq. (3.38) as

$$\frac{1}{k_{\text{Overall}}} = \frac{k_{\text{Rxn}} + k_G a_G}{k_G a_G k_{\text{Rxn}}} + \left(\frac{1}{k_L}\right)\frac{1}{\text{RPM}} \tag{3.42}$$

Thus a plot of $1/k_{\text{Overall}}$ versus 1/RPM yields a straight line with $(k_{\text{Rxn}} + k_G a_G)/k_G a_G k_{\text{Rxn}}$ as its intercept and $1/k_L$ as its slope. Fig. 3.4 shows such a plot. If we draw a horizontal line through the intercept,

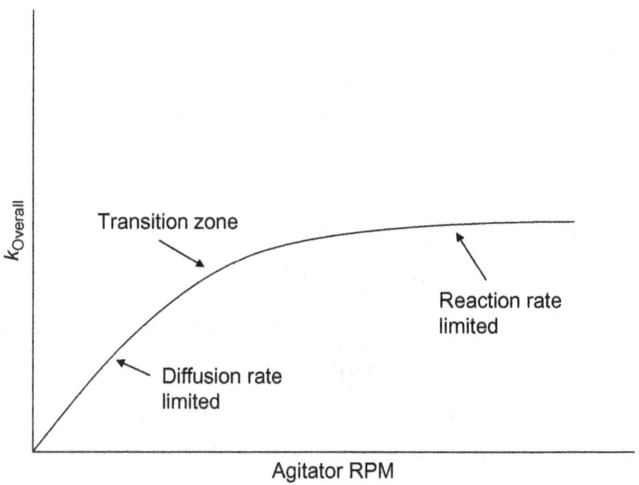

Figure 3.3 $k_{Overall}$ *as a function of agitator RPMs.*

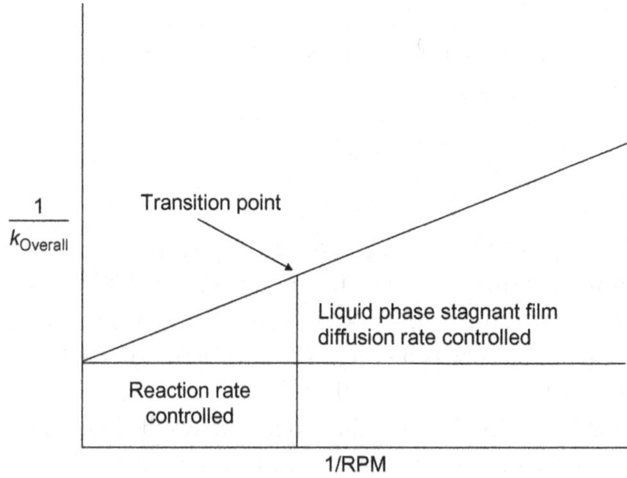

Figure 3.4 Schematic representation of a gas–liquid semibatch process.

then the area below it represents the combination of reaction rate and gas phase diffusion rate control. Of course, if the gas is pure or the liquid essentially not volatile, then there is no gas diffusion and this area simply represents reaction rate control. The line with slope k_L represents the ever increasing importance of dissolved gas diffusion through the stagnant liquid film surrounding each gas bubble in the semibatch reactor. If we draw a vertical line and move it along the horizontal axis, at some point, we will find that half the line is in the reaction rate controlled area and the other half of the line is above the reaction rate controlled area. We call this point the transition from reaction rate control to liquid phase diffusion control. To the right of this line, the semibatch process is liquid phase diffusion rate controlled; to the left of this line, the semibatch process is reaction rate controlled. When scaling a semibatch process, we must know whether the model is to the right or left of this line to ensure that the prototype operates similarly to it.

CONTROL REGIMES FOR IMMISCIBLE LIQUID–LIQUID SEMIBATCH PROCESSES

Fig. 3.5 represents a semibatch reactor process involving immiscible liquids. One liquid is completely charged to the reactor: it is the bulk reactant BR. The second liquid "sprays" into the reactor just above

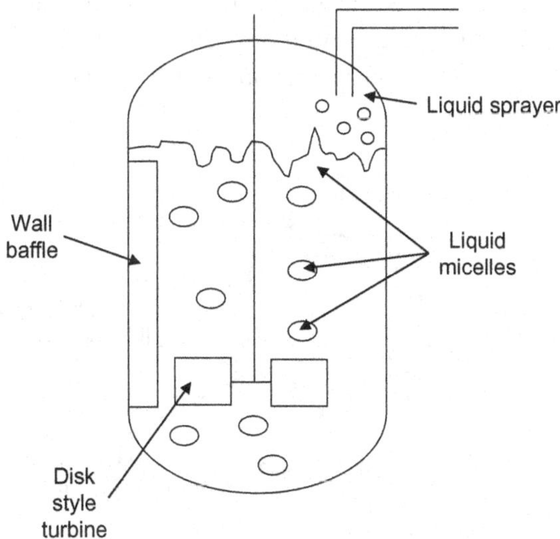

Figure 3.5 Macroscopic features of immiscible liquids semibatch process.

the surface of the bulk reactant. We call this second liquid the limiting reactant or limiting reagent LR. LR forms micelles upon contacting BR. The agitator distributes these micelles throughout BR. The baffles assist this distribution by preventing the formation of a vortex in BR.

Fig. 3.6 shows the microscopic features of this process. Each LR micelle contains pure LR liquid. If LR solubilizes small amounts of BR, then a stagnant molecular film exists adjacent to the liquid–liquid interface. If LR is pure and if it does not solubilize BR, then no stagnant film forms inside the LR micelle. For this analysis, we assume a stagnant film forms inside the LR micelle [1]. Diffusion governs the movement of liquid LR molecules within this stagnant film. Physical equilibrium, described by a distribution coefficient, controls the movement of LR across this interface and into BR. A stagnant film of BR surrounds each LR micelle and diffusion governs the movement of solubilized LR within it. BR comprises the bulk fluid phase moving between each LR micelle.

We describe, mathematically, the rate of LR movement through the stagnant LR film as

$$R_{\mathrm{LRF}} = k_{\mathrm{LRF}} a_{\mathrm{LRF}} ([\mathrm{LR}]_{\mathrm{Feed}} - [\mathrm{LR}]_{\mathrm{Interface/LR}}) \tag{3.43}$$

where R_{LRF} is the rate of LR molecules moving across the stagnant film of LR inside the micelle (moles/m$^3\cdot$s); k_{LRF} is the rate at which LR molecules move across the stagnant film of LR (m/s); a_{LRF} is the surface area to volume of the stagnant LR film inside the LR micelle (m^2/m^3); $[\mathrm{LR}]_{\mathrm{Feed}}$ is the concentration of LR entering the semibatch

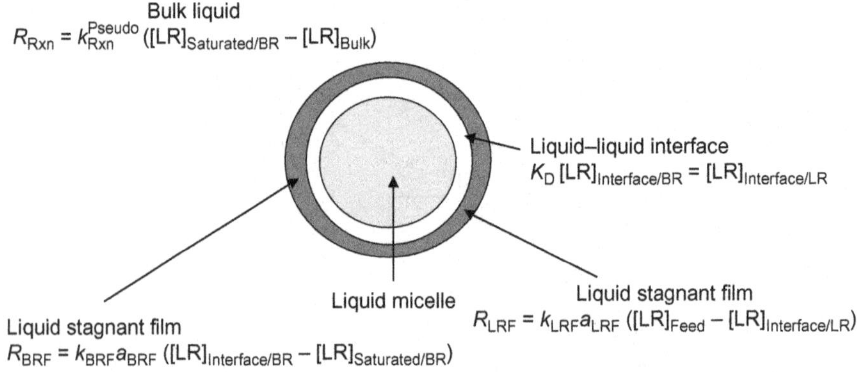

Figure 3.6 Microscopic features of liquid–liquid or immiscible liquids semibatch process.

reactor (mole/m^3); and $[LR]_{Interface/LR}$ is the concentration of LR at the liquid–liquid interface of the micelle (mole/m^3).

The distribution coefficient K_D is the concentration of LR in LR at the liquid–liquid interface and the concentration of LR in BR at the liquid–liquid interface [4,5]; thus mathematically, it is

$$K_D = \frac{[LR]_{Interface/LR}}{[LR]_{Interface/BR}} \quad (3.44)$$

where $[LR]_{Interface/BR}$ is the concentration of LR in BR at the liquid–liquid interface of the micelle (mole/m^3). Rearranging Eq. (3.44) gives

$$K_D[LR]_{Interface/BR} = [LR]_{Interface/LR} \quad (3.45)$$

The rate at which dissolved LR moves through the stagnant BR film surrounding each micelle is

$$R_{BRF} = k_{BRF}a_{BRF}([LR]_{Interface/BR} - [LR]_{Saturated/BR}) \quad (3.46)$$

where R_{BRF} is the rate of LR molecules moving across the stagnant film of BR surrounding the micelle (moles/m$^3 \cdot$ s); k_{BRF} is the rate at which LR molecules move across the stagnant film of BR (m/s); a_{BRF} is the surface area to volume of the stagnant film of BR outside the LR micelle (m^2/m^3); $[LR]_{Interface/BR}$ is the concentration of LR at the liquid–liquid interface of the micelle (mole/m^3); and $[LR]_{Saturated/BR}$ is the saturated concentration of LR in BR (mole/m^3).

If the chemical reaction is irreversible, then the chemical reaction within BR is

$$R_{Rxn} = k_{Rxn}^{Pseudo}([LR]_{Saturated/BR} - [LR]_{Bulk}) \quad (3.47)$$

We write R_{Rxn} as a first-order reaction with respect to LR because, for this process

$$[BR] \gg [LR] \quad (3.48)$$

Therefore

$$k_{Rxn}^{Pseudo} = k_{Rxn}[BR] \quad (3.49)$$

Unfortunately, in the above rate equations, we only know $[LR]_{Feed}$ and $[LR]_{Bulk}$, which we measure by properly sampling the contents of the semibatch reactor and using the appropriate analytical methods. We must, then, reduce the above set of rate equations to one equation containing $[\mathrm{LR}]_{\mathrm{Feed}}$ and $[\mathrm{LR}]_{\mathrm{Bulk}}$.

Substituting the distribution coefficient into Eq. (3.43) gives

$$R_{\mathrm{LRF}} = k_{\mathrm{LRF}} a_{\mathrm{LRF}}([\mathrm{LR}]_{\mathrm{Feed}} - K_{\mathrm{D}}[\mathrm{LR}]_{\mathrm{Interface/BR}}) \tag{3.50}$$

then solving for $[\mathrm{LR}]_{\mathrm{Interface/BR}}$ yields

$$\frac{R_{\mathrm{LRF}}}{k_{\mathrm{LRF}} a_{\mathrm{LRF}}} - [\mathrm{LR}]_{\mathrm{Feed}} = -K_{\mathrm{D}}[\mathrm{LR}]_{\mathrm{Interface/BR}} \tag{3.51}$$

or

$$\frac{R_{\mathrm{LRF}}}{K_{\mathrm{D}} k_{\mathrm{LRF}} a_{\mathrm{LRF}}} - \frac{[\mathrm{LR}]_{\mathrm{Feed}}}{K_{\mathrm{D}}} = -[\mathrm{LR}]_{\mathrm{Interface/BR}} \tag{3.52}$$

Substituting Eq. (3.52) into

$$R_{\mathrm{BRF}} = k_{\mathrm{BRF}} a_{\mathrm{BRF}}([\mathrm{LR}]_{\mathrm{Interface/BR}} - [\mathrm{LR}]_{\mathrm{Saturated/BR}}) \tag{3.53}$$

gives

$$R_{\mathrm{BRF}} = k_{\mathrm{BRF}} a_{\mathrm{BRF}}\left(-\frac{R_{\mathrm{LRF}}}{K_{\mathrm{D}} k_{\mathrm{LRF}} a_{\mathrm{LRF}}} + \frac{[\mathrm{LR}]_{\mathrm{Feed}}}{K_{\mathrm{D}}} - [\mathrm{LR}]_{\mathrm{Saturated/BR}}\right) \tag{3.54}$$

Solving for $[\mathrm{LR}]_{\mathrm{Saturated/BR}}$ yields

$$\frac{R_{\mathrm{BRF}}}{k_{\mathrm{BRF}} a_{\mathrm{BRF}}} + \frac{R_{\mathrm{LRF}}}{K_{\mathrm{D}} k_{\mathrm{LRF}} a_{\mathrm{LRF}}} - \frac{[\mathrm{LR}]_{\mathrm{Feed}}}{K_{\mathrm{D}}} = -[\mathrm{LR}]_{\mathrm{Saturated/BR}} \tag{3.55}$$

Substituting Eq. (3.55) into

$$R_{\mathrm{Rxn}} = k_{\mathrm{Rxn}}^{\mathrm{Pseudo}}([\mathrm{LR}]_{\mathrm{Saturated/BR}} - [\mathrm{LR}]_{\mathrm{Bulk}}) \tag{3.56}$$

gives us, after rearranging terms

$$\frac{R_{\mathrm{LRF}}/K_{\mathrm{D}}}{k_{\mathrm{LRF}} a_{\mathrm{LRF}}} + \frac{R_{\mathrm{BRF}}}{k_{\mathrm{BRF}} a_{\mathrm{BRF}}} + \frac{R_{\mathrm{Rxn}}}{k_{\mathrm{Rxn}}^{\mathrm{Pseudo}}} = \left(\frac{[\mathrm{LR}]_{\mathrm{Feed}}}{K_{\mathrm{D}}} - [\mathrm{LR}]_{\mathrm{Bulk}}\right) \tag{3.57}$$

We now have all the rates related to the two variables we can measure. We know the overall rate of the process is

$$R_{\mathrm{Overall}} = k_{\mathrm{Overall}}\left(\frac{[\mathrm{LR}]_{\mathrm{Feed}}}{K_{\mathrm{D}}} - [\mathrm{LR}]_{\mathrm{Bulk}}\right) \tag{3.58}$$

where R_{Overall} is the overall rate for the semibatch process (moles/$\text{m}^3 \cdot \text{s}$) and k_{Overall} is the overall rate constant for the semibatch process (s^{-1}). Since none of the rates can be faster than the slowest rate and the overall rate equals the slowest rate, we can write

$$\frac{R_{\text{LRF}}}{K_{\text{D}}} = R_{\text{BRF}} = R_{\text{Rxn}} = R_{\text{Overall}} \tag{3.59}$$

Therefore

$$\frac{R_{\text{Overall}}}{k_{\text{LRF}}a_{\text{LRF}}} + \frac{R_{\text{Overall}}}{k_{\text{BRF}}a_{\text{BRF}}} + \frac{R_{\text{Overall}}}{k_{\text{Rxn}}^{\text{Pseudo}}} = \left(\frac{[\text{LR}]_{\text{Feed}}}{K_{\text{D}}} - [\text{LR}]_{\text{Bulk}}\right) \tag{3.60}$$

or

$$R_{\text{Overall}}\left(\frac{1}{k_{\text{LRF}}a_{\text{LRF}}} + \frac{1}{k_{\text{BRF}}a_{\text{BRF}}} + \frac{1}{k_{\text{Rxn}}^{\text{Pseudo}}}\right) = \left(\frac{[\text{LR}]_{\text{Feed}}}{K_{\text{D}}} - [\text{LR}]_{\text{Bulk}}\right) \tag{3.61}$$

Rearranging gives us

$$R_{\text{Overall}} = \left(\frac{1}{\dfrac{1}{k_{\text{LRF}}a_{\text{LRF}}} + \dfrac{1}{k_{\text{BRF}}a_{\text{BRF}}} + \dfrac{1}{k_{\text{Rxn}}^{\text{Pseudo}}}}\right)\left(\frac{[\text{LR}]_{\text{Feed}}}{K_{\text{D}}} - [\text{LR}]_{\text{Bulk}}\right) \tag{3.62}$$

Thus

$$k_{\text{Overall}} = \left(\frac{1}{\dfrac{1}{k_{\text{LRF}}a_{\text{LRF}}} + \dfrac{1}{k_{\text{BRF}}a_{\text{BRF}}} + \dfrac{1}{k_{\text{Rxn}}^{\text{Pseudo}}}}\right) \tag{3.63}$$

which becomes

$$\frac{1}{k_{\text{Overall}}} = \frac{1}{k_{\text{LRF}}a_{\text{LRF}}} + \frac{1}{k_{\text{BRF}}a_{\text{BRF}}} + \frac{1}{k_{\text{Rxn}}^{\text{Pseudo}}} \tag{3.64}$$

We generally interpret Eq. (3.64) as representing a series of resistances. The overall resistance, $1/k_{\text{Overall}}$, equals the sum of the resistance to LR diffusion across the stagnant liquid film inside the micelle, the resistance to LR diffusion across the liquid film surrounding the micelle, and the resistance to LR reaction. Note that $k_{\text{Rxn}}^{\text{Pseudo}}$, k_{BRF}, and

k_{LRF} are constants. $k_{\mathrm{Rxn}}^{\mathrm{Pseudo}}$ describes the rate of the chemical reaction per unit time; that is, under a given set of operating conditions, for example, temperature and pressure, $k_{\mathrm{Rxn}}^{\mathrm{Pseudo}}$ does not change. k_{BRF} and k_{LRF} depend upon the mean free path of a molecule in the liquid and the mean free path depends upon the liquid density, which, in our case, we assume is constant. Therefore, any fluctuations in k_{Overall} must arise from fluctuations in a_{BRF} or a_{LRF} or both.

If the micelles of a given liquid–liquid semibatch process are spherical, then

$$a_{\mathrm{LRF}} \text{ or } a_{\mathrm{BRF}} = \frac{4\pi r^2}{(4/3)\pi r^3} = \frac{3}{r} \tag{3.65}$$

where r is the radius of the micelle. So, as the micelle shrinks due to solubilization and chemical reaction, a_{BRF} and a_{LRF} increase. An increase in a_{BRF} and a_{LRF} leads directly to an increase in k_{Overall} since

$$\frac{1}{k_{\mathrm{Overall}}} \propto \frac{1}{a_{\mathrm{LRF}}} \text{and} \frac{1}{a_{\mathrm{BRF}}} \tag{3.66}$$

k_{Overall} increases because, as the LR and BR stagnant film thicknesses decrease, the time to cross each of them shortens due to k_{BRF} and k_{LRF} being constant.

a_{BRF} depends not only on micelle radius but also on the velocity of BR streaming past an LR micelle. For a constant micelle radius, we obtain the thinnest BR stagnant film at the highest possible agitation rate. To further reduce BR stagnant film thickness, we must decrease the radius of the LR micelle. If we operate a semibatch process at an agitation rate that minimizes the BR stagnant film thickness with respect to BR fluid velocity, then we can represent the process schematically as shown in Fig. 3.7, where we plot $1/k_{\mathrm{Overall}}$ as a function of micelle radius r. Since resistance to LR chemical reaction, that is, $1/k_{\mathrm{Rxn}}^{\mathrm{Pseudo}}$, is independent of micelle radius, it plots as a horizontal line in Fig. 3.7. Resistance to LR diffusion across the BR stagnant film surrounding an LR micelle plots in Fig. 3.7 as a straight line with slope $1/k_{\mathrm{BRF}}$ while resistance to LR diffusion across the LR stagnant film inside the LR micelle plots as a straight line with slope $1/k_{\mathrm{LRF}}$. Note the vertical line at $r_{\mathrm{Transition}}$ in Fig. 3.7: the length of $1/k_{\mathrm{Overall}}$ below the horizontal line representing resistance to LR chemical reaction equals the length of $1/k_{\mathrm{Overall}}$ from that same horizontal line to the

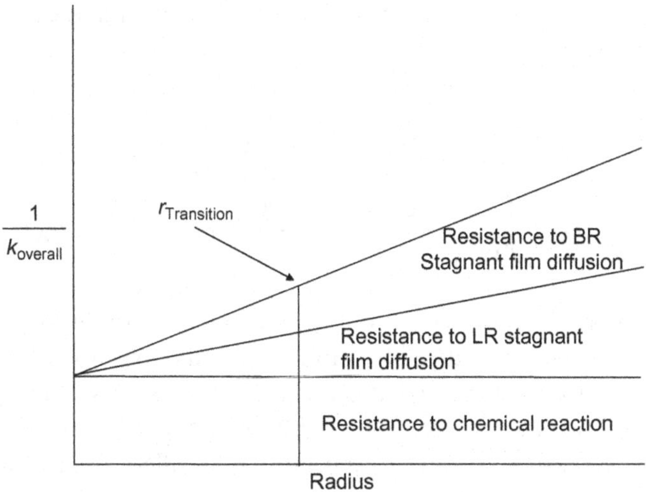

Figure 3.7 Schematic representation of resistance to reactant consumption in semibatch processes.

topmost straight line. In other words, resistance to LR chemical reaction equals resistance to diffusion across both stagnant film layers. To the right of $r_{Transition}$, the semibatch process is film diffusion rate limited; to the left of $r_{Transition}$, the semibatch process is reaction rate limited. We want to operate the semibatch process to the left of $r_{Transition}$. Therefore, we will operate the semibatch process at the highest possible agitation rate and the smallest possible micelle radius.

This example shows that an immiscible liquid–liquid semibatch process can be either reaction rate controlled, limiting reagent diffusion rate controlled, or bulk reactant diffusion rate controlled. Before scaling such a process, we must determine which regime controls the process and decide whether to accept that result or to change to a more profitable controlling regime. If we do not know which regime controls the process, then our scaling effort may produce results that surprise us.

CONTROL REGIMES FOR FIXED-BED REACTORS

Fixed-bed reactors come in all sizes, but their shape is generally restricted to that of a circular cylinder which is later filled with solid-supported catalyst. Feed enters one end of the reactor and product exits the other end of the reactor. The component mass balance for a fixed-bed reactor is

$$\frac{\partial C_A}{\partial t} + \left(v_r \frac{\partial C_A}{\partial r} + \frac{v_\theta}{r} \frac{\partial C_A}{\partial \theta} + v_z \frac{\partial C_A}{\partial z} \right) = D_{AB} \left(\frac{1}{r} \frac{\partial}{\partial r} \left(r \frac{\partial C_A}{\partial r} \right) + \frac{1}{r^2} \frac{\partial^2 C_A}{\partial \theta^2} + \frac{\partial^2 C_A}{\partial z^2} \right) + R_A \tag{3.67}$$

where C_A is the concentration of component A in moles/m^3 [NL^{-3}]—note, the brackets indicate dimension, not units. We operate fixed-bed reactors at constant feed and product flow rates; thus they operate at steady state, which means $\partial C_A/\partial t = 0$. At high feed rates, the flow through a fixed-bed reactor, especially one filled with small solid-supported catalyst pellets or extrudates, behaves in a plug flow manner. Therefore, $v_r = v_\theta = 0$. The reactor design, catalyst size, and fluid flow rate combine to determine whether the dispersion terms impact the performance of a fixed-bed reactor; the dispersion terms are

$$\frac{1}{r} \frac{\partial}{\partial r} \left(r \frac{\partial C_A}{\partial r} \right) + \frac{1}{r^2} \frac{\partial^2 C_A}{\partial \theta^2} + \frac{\partial^2 C_A}{\partial z^2} \tag{3.68}$$

The Peclet number for the process and the various aspect ratios of the fixed-bed reactor determine the impact of the dispersion terms. The mass Peclet number quantifies the ratio of bulk mass transport to diffusive mass transport. We define the mass Peclet number as

$$Pe = \frac{Lv}{D_{AB}} \tag{3.69}$$

where L is a characteristic length [L]; v is fluid velocity [LT^{-1}]; and D_{AB} the diffusivity constant for component A in bulk component B [L^{2}T^{-1}]. L can be the height of the catalyst mass, identified as Z, or the diameter of the reactor, identified as D, both, dimensionally, are [L]. The product of the directional aspect ratio and the Peclet number determines the importance of dispersion in that direction. If $(Z/D_p) \cdot Pe$ is large, then dispersion in the axial direction, that is, along the z-axis, is negligible. D_p is the diameter of the solid-supported catalyst [L]. Generally, Z is large for a fixed-bed reactor and D_p is small; thus we can confidently neglect the $\partial^2 C_A/\partial z^2$ dispersion term. With regard to the axial aspect ratio itself, Carberry purports that $Z/D_p \geq 150$ ensures no axial dispersion in a fixed-bed reactor [6].

In the radial direction, we seek the opposite outcome because conduction is the only mass transfer mechanism in that direction. Thus we want a small radial Pe number, which is Dv/D_{AB}. It is heat transfer at the fixed-bed reactor wall that induces radial conduction and mass transfer. Hence, radial dispersion can only be neglected for adiabatic reactors. However, Carberry purports that radial aspect ratios R/D_p of 3 to 4 ensure negligible radial dispersion, where R is the reactor's radius [L] [7]. If we meet this criterion, then $\partial C_A/\partial r$ is zero. Chemical engineers generally neglect dispersion in the azimuthal direction. Therefore, the above component mass balance reduces to

$$v_z \frac{dC_A}{dz} = R_A \tag{3.70}$$

where R_A represents reactant consumption or product formation.

We generally do not characterize fixed-bed reactor operation by its fluid velocity. Instead, we characterize fixed-bed operation by its volumetric flow rate. Rearranging Eq. (3.70), then multiplying by the cross-sectional area of the reactor, we obtain

$$v_z A\, dC_A = R_A A\, dz \tag{3.71}$$

Note that $v_z A$ is volumetric flow rate [L^3T^{-1}] and $A\,dz$ is dV, the differential volume of catalyst [L^3]. V is the fluid volume of the catalyst mass, which is the true volume of the reactor. However, the chemical processing industry generally identifies V as the weight W of the solid-supported catalyst loaded in the reactor. Using W instead of V complicates our analysis for a fixed-bed reactor. Therefore, we will identify V as the true volume for a fixed-bed reactor. That volume is

$$\frac{W_{Catalyst}}{\rho_{LBD}} = V_{Catalyst} \tag{3.72}$$

where $W_{Catalyst}$ is the weight of catalyst loaded in the reactor [M] and ρ_{LBD} is the loose bulk density of the catalyst [ML^{-3}]. We use loose bulk density because we do not shake a commercial reactor while filling it. We may shake a laboratory reactor while filling it, but we generally do not shake commercial reactors when filling them. Note that $V_{Catalyst}$ includes fluid volume as well as solid volume. We are interested in the fluid volume of the catalyst mass, which is

$$V_{Fluid} = \varepsilon V_{Catalyst} \tag{3.73}$$

where ε is the void fraction of the solid-supported catalyst.

Thus the above mass balance becomes

$$Q\, dC_{\mathrm{A}} = R_{\mathrm{A}}\, dV_{\mathrm{Fluid}} \tag{3.74}$$

Rearranging Eq. (3.74) gives us

$$\frac{dC_{\mathrm{A}}}{R_{\mathrm{A}}} = \frac{dV_{\mathrm{Fluid}}}{Q} \tag{3.75}$$

The boundary conditions for Eq. (3.75) are $C_{\mathrm{A}} = C_{\mathrm{A,In}}$ at $V = 0$ and $C_{\mathrm{A}} = C_{\mathrm{A,Out}}$ at $V = V_{\mathrm{Fluid}}$. Integrating with these boundary conditions yields

$$\int_{C_{A,\mathrm{In}}}^{C_{A,\mathrm{Out}}} \frac{dC_{\mathrm{A}}}{R_{\mathrm{A}}} = \int_0^{V_{\mathrm{Fluid}}} \frac{dV_{\mathrm{Fluid}}}{Q} = \frac{V_{\mathrm{Fluid}}}{Q} \tag{3.76}$$

V_{Fluid}/Q has units of time [T], generally in minutes or seconds. We call V_{Fluid}/Q space time and it represents the average time required to traverse a given flow path from the leading edge of the catalyst mass to the trailing edge of the catalyst mass. We cannot as yet integrate the left-hand side of Eq. (3.76) because we have not specified R_{A} mathematically. To do so, we must know what happens within the catalyst mass during reactor operation.

But, just what does occur within the catalyst mass during reactor operation?

First, fluid enters one of the myriad flow tubes or channels passing through the catalyst mass. These flow channels form between catalyst pellets or extrudates. The bulk fluid moves through the catalyst mass via these flow tubes or channels.

Second, whenever a fluid flows over a solid or liquid surface, a stagnant film forms along that surface. At the surface, fluid velocity in the direction of bulk flow is zero. At the outer edge of this stagnant film, fluid velocity is that of the bulk fluid. Within the stagnant film, a velocity gradient in the direction of bulk fluid flow exists. However, there is no fluid velocity component normal, that is, perpendicular to the solid or liquid surface. In other words, convection does not occur across the stagnant film. Thus reactant and product molecules diffuse across the stagnant film.

Third, most solid-supported catalysts are porous, which greatly increases their surface area. Increasing surface area leads to an increased number of catalytic sites available for reaction, thereby increasing catalyst productivity. Reactant molecules migrate along these pores via diffusion. When they encounter an empty catalyst site, they become product. These product molecules must then diffuse through the pore network of the solid-supported catalyst, diffuse across the stagnant film surrounding each catalyst pellet, then enter the bulk fluid to exit the catalyst mass. The concentration difference between a feed sample and a product sample represents the sum of these mechanisms.

Fig. 3.8 presents a schematic of catalysis in a porous solid. The reactant concentration at the outer boundary of the stagnant film is the concentration of reactant in the feed, that is, it is C_{BF} in moles/m^3. Reactant molecules move across the stagnant film by molecular diffusion, which we generally model as a linear concentration difference. That difference is $C_{BF} - C_{SF}$, where C_{SF} is the concentration of reactant at the surface of the catalyst in moles/m^3. Reactant then diffuses from the surface of the catalyst along pores to the catalytic sites inside the solid. Reactant movement within the pore is also by molecular diffusion, which we model as a linear concentration difference. Catalytic sites occur along the length of the pore, thus reactant concentration

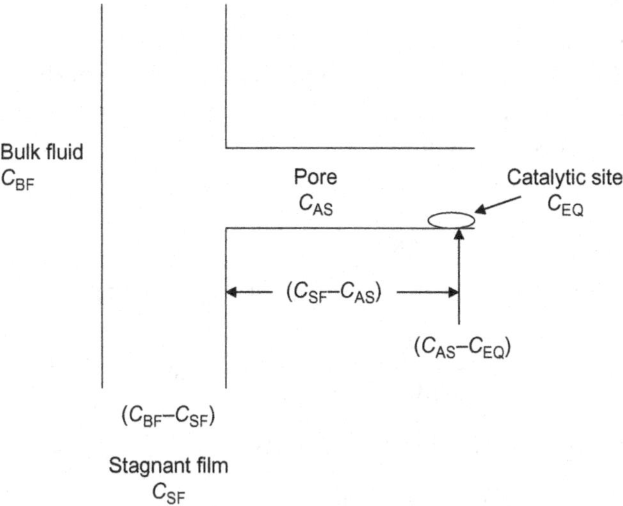

Figure 3.8 Schematic of adsorption impedances.

changes along a pore's length. Reactant concentration at a given catalyst site is C_{AS} (moles/m^3). Thus the concentration difference to that point in the pore is $C_{SF} - C_{AS}$. Equilibrium may be established at the catalytic site; equilibrium concentration is C_{Eq} (moles/m^3).

We can write the rate of each of the above-described mechanisms as a concentration difference. The reactant conversion rate is

$$R_{Rxn} = k_{Rxn}(C_{AS} - C_{Eq}) \tag{3.77}$$

where k_{Rxn} is the reaction rate constant at the catalytic site. k_{Rxn} has units of min^{-1} or s^{-1}. The rate of reactant movement along the pore is

$$R_{PD} = k_{PD}\left(\frac{A_P}{V_P}\right)(C_{SF} - C_{AS}) \tag{3.78}$$

where k_{PD} is the rate at which molecules move along the pores (m/s); A_P is the average cross-sectional area of a pore (m^2); and V_P is the average pore volume (m^3). The rate of reactant movement through the stagnant film surrounding the catalyst pellet is

$$R_{SFD} = k_{SFD}\left(\frac{S_{Film}}{V_{Film}}\right)(C_{BF} - C_{SF}) \tag{3.79}$$

where k_{SFD} is the rate at which molecules move across the stagnant film (m/s), S_{Film} is the surface area (m^2), and V_{Film} is the volume (m^3) of the stagnant film surrounding the catalyst pellet.

The only reactant concentrations known with any accuracy are C_{BF} and C_{Eq}. Thus the mathematical expression for the overall rate of reactant conversion must be in terms of C_{BF} and C_{Eq}. Solving Eq. (3.77) for R_{Rxn} in terms of C_{AS} yields

$$\frac{R_{Rxn}}{k_{Rxn}} + C_{Eq} = C_{AS} \tag{3.80}$$

Solving Eq. (3.78) for R_{PD} in terms of C_{SF} provides

$$\frac{R_{PD}}{k_{PD}(A_P/V_P)} + C_{AS} = C_{SF} \tag{3.81}$$

Substituting C_{AS} into Eq. (3.81) gives us

$$\frac{R_{Rxn}}{k_{Rxn}} + \frac{R_{PD}}{k_{PD}(A_P/V_P)} + C_{Eq} = C_{SF} \tag{3.82}$$

Solving Eq. (3.79) for R_{SFD} in terms of C_{SF} yields

$$\frac{R_{SFD}}{k_{SFD}(S_{Film}/V_{Film})} + C_{BF} = C_{SF} \tag{3.83}$$

then substituting C_{SF} from Eq. (3.83) into Eq. (3.82) produces

$$\frac{R_{Rxn}}{k_{Rxn}} + \frac{R_{PD}}{k_{PD}(A_P/V_P)} + \frac{R_{SFD}}{k_{SFD}(S_{Film}/V_{Film})} + C_{Eq} = C_{BF} \tag{3.84}$$

Rearranging this result expresses the rate of reactant conversion in terms of C_{BF} and C_{Eq}. Thus

$$\frac{R_{Rxn}}{k_{Rxn}} + \frac{R_{PD}}{k_{PD}(A_P/V_P)} + \frac{R_{SFD}}{k_{SFD}(S_{Film}/V_{Film})} = C_{BF} - C_{Eq} \tag{3.85}$$

By assuming $R_{Rxn} = R_{PD} = R_{SFD} = R$, then rearranging Eq. (3.85), the overall rate of reactant conversion in terms of C_{BF} and C_{Eq} becomes

$$R\left\{\frac{1}{k_{Rxn}} + \frac{1}{k_{PD}(A_P/V_P)} + \frac{1}{k_{SFD}(S_{Film}/V_{Film})}\right\} = C_{BF} - C_{Eq} \tag{3.86}$$

or

$$R = \left\{\frac{1}{\dfrac{1}{k_{Rxn}} + \dfrac{1}{k_{PD}(A_P/V_P)} + \dfrac{1}{k_{SFD}(S_{Film}/V_{Film})}}\right\} C_{BF} - C_{Eq} \tag{3.87}$$

The overall reaction rate constant is, by definition

$$k_{Overall} = \left\{\frac{1}{\dfrac{1}{k_{Rxn}} + \dfrac{1}{k_{PD}(A_P/V_P)} + \dfrac{1}{k_{SFD}(S_{Film}/V_{Film})}}\right\} \tag{3.88}$$

Inverting $k_{Overall}$ yields

$$\frac{1}{k_{Overall}} = \frac{1}{k_{Rxn}} + \frac{1}{k_{PD}(A_P/V_P)} + \frac{1}{k_{SFD}(S_{Film}/V_{Film})} \tag{3.89}$$

which is, again, an equation representing a set of resistances in series.

We can simplify Eq. (3.89) by combining $1/k_{Rxn}$ and $1/k_{PD}(A_P/V_P)$. Doing so yields

$$\frac{1}{k_{Overall}} = \frac{k_{PD}(A_P/V_P) + k_{Rxn}}{k_{PD}(A_P/V_P)k_{Rxn}} + \frac{1}{k_{SFD}(S_{Film}/V_{Film})} \tag{3.90}$$

which reduces to

$$\frac{1}{k_{\text{Overall}}} = \frac{1}{\eta k_{\text{Rxn}}} + \frac{1}{k_{\text{SFD}}(S_{\text{Film}}/V_{\text{Film}})} \tag{3.91}$$

where η is defined as

$$\eta = \frac{k_{\text{PD}}(A_{\text{P}}/V_{\text{P}})}{k_{\text{PD}}(A_{\text{P}}/V_{\text{P}}) + k_{\text{Rxn}}} \tag{3.92}$$

We call η the "effectiveness factor." The effectiveness factor accounts for the concentration difference along the pore of a solid-supported catalyst. If $k_{\text{PD}}(A_{\text{P}}/V_{\text{P}}) \gg k_{\text{Rxn}}$, that is, if the fixed-bed reactor is reaction rate limited, then $\eta = 1$ and if $k_{\text{Rxn}} \gg k_{\text{PD}}(A_{\text{P}}/V_{\text{P}})$, that is, if the fixed-bed reactor is pore diffusion rate limited, then $\eta < 1$. η depends only upon the pore structure of the solid-supported catalyst and is readily calculated via a variety of published methods [8].

We can now specify R_{A} and substitute it into the mass balance

$$\int_{C_{\text{A,In}}}^{C_{\text{A,Out}}} \frac{dC_{\text{A}}}{R_{\text{A}}} = \int_0^V \frac{dV_{\text{Fluid}}}{Q} = \frac{V_{\text{Fluid}}}{Q} \tag{3.93}$$

and solve for C_{A}.

Commercial plants operate at high volumetric flow rates because their purpose is to produce as much product per unit time as possible. Laboratory and pilot plant facilities have a different purpose. Their purpose is to produce quality information that can be used to design a new commercial process or to support the operations of an existing commercial facility. We design commercial facilities to store large volumes of feed and product and to circulate large quantities of process fluids safely and with minimum environmental impact. We operate laboratory and pilot plant reactors at low volumetric flow rates because we want to minimize the volume of feed stored at the research facility. Another reason we operate laboratory and pilot plant reactors at low volumetric flow rate is waste disposal: the product produced by such reactors cannot be sold. Therefore, it must be disposed, which is expensive. Also, laboratory and pilot plant processes circulate small quantities of process fluids so that any leak is small, thereby minimizing the safety and environmental issues arising from the leak. Finally, if we are developing an entirely new process, we will be unsure of all the potential hazards it entails. Thus we will keep all reactive volumes small to reduce the impact of a runaway reaction. In other words, laboratory and pilot plant fixed-bed reactors generally operate in the laminar flow regime while commercial-sized fixed-bed reactors operate in the turbulent flow regime.

This flow regime difference can have a dramatic impact upon the results produced by a laboratory or pilot plant fixed-bed reactor. In the laminar flow regime, the stagnant film surrounding each solid-supported catalyst pellet will be much thicker than the stagnant film surrounding solid-supported catalyst in a commercial reactor operating in the turbulent flow regime [9]. The diffusion rate constant, expressed as m/s, is the same in both flow regimes; however, the time to cross the stagnant film is larger for laminar flow than for turbulent flow since the stagnant film is thicker for laminar flow than for turbulent flow. Thus product formation will be slower in the laminar flow regime than in the turbulent flow regime.

If the laboratory and pilot plant fixed-bed reactors are stagnant film diffusion rate limited, then k_{Overall} will plot as scatter around an average value, which may or may not be recognized as k_{SFD}, the stagnant film diffusion rate constant. Not being aware of this potentiality produces highly expensive, inconclusive process support efforts and catalyst development programs. Years can be spent evaluating different catalyst sizes, shapes, and compositions to no avail because all the data scatters about one value. Unbeknownst to those working on the project, that point is the stagnant film diffusion rate constant. Eventually, the project will be canceled due to no conclusive results.

A similar situation can arise for pore diffusion rate-limited processes. In this situation the physical structure of the solid support must be altered to improve catalyst performance, that is, to increase k_{Overall}. Changing the chemical composition of the solid-supported catalyst will not alter k_{Overall}. All the data produced by the laboratory and pilot plant reactors will scatter in one region of the plot, which is the pore diffusion rate constant for the process.

If a solid-supported catalyst is reaction rate limited, then changing size, shape, or pore structure will not improve catalyst performance. In this case, the only way to improve catalyst performance is to alter the chemical composition of the catalyst.

The message is to know the flow regime occurring in a fixed-bed reactor and know the rate controlling step of a solid-supported catalyzed process before attempting to improve a catalyzed process or scale a catalyzed process.

Consider the modified resistance equation (Eq. 3.91), which is

$$\frac{1}{k_{\text{Overall}}} = \frac{1}{\eta k_{\text{Rxn}}} + \frac{1}{k_{\text{SFD}}(S_{\text{Film}}/V_{\text{Film}})}$$

We can determine k_{Overall} directly by measuring $C_{\text{BF,In}}$ and $C_{\text{BF,Out}}$, then calculating k_{Overall} from the following equation:

$$R = k_{\text{Overall}}(C_{\text{BF,Out}} - C_{\text{BF,In}}) \tag{3.95}$$

We can determine the value of k_{Rxn} from laboratory experiments that obviate any diffusion effects. What we do not know is k_{SFD} $(S_{\text{Film}}/V_{\text{Film}})$. However, we do know that k_{SFD} is constant at a given reactor operating temperature and pressure. Thus only $S_{\text{Film}}/V_{\text{Film}}$ responds to changes in the velocity of the bulk fluid over the surface of the catalyst because $S_{\text{Film}}/V_{\text{Film}}$ is the ratio of the stagnant film's surface area to its volume; therefore, it is inversely proportional to the stagnant film's thickness. Mathematically

$$\frac{S_{\text{Film}}}{V_{\text{Film}}} \propto \frac{1}{\delta} \tag{3.96}$$

where δ is the thickness of the stagnant film surrounding each solid-supported catalyst pellet. But, stagnant film thickness is proportional to the Reynolds number of the bulk fluid, that is [3]

$$\delta \propto \frac{1}{\sqrt{Re_z}} = \frac{1}{\sqrt{\rho D v_z/\mu}} = \sqrt{\frac{\mu}{\rho D v_z}} \tag{3.97}$$

Thus as the interstitial, linear velocity of the bulk fluid increases, δ decreases. And, as δ decreases, reactant and product molecules spend less time traversing the stagnant film surrounding each solid-supported catalyst pellet, which increases k_{Overall}. We can relate δ to v_z since the fluid density, fluid viscosity, and reactor diameter remain constant during the velocity change. Thus

$$\frac{S_{\text{Film}}}{V_{\text{Film}}} \propto f(v_z) \tag{3.98}$$

where $f(v_z)$ represents an unspecified function of v_z. The modified resistance equation (Eq. 3.91) can now be written as

$$\frac{1}{k_{\text{Overall}}} = \frac{1}{\eta k_{\text{Rxn}}} + \frac{1}{k_{\text{SFD}} * f(v_z)} \tag{3.99}$$

which has the form of a straight line if we plot $1/k_{\text{Overall}}$ as a function of $1/f(v_z)$. The slope is $1/k_{\text{SFD}}$ and the intercept is $1/\eta k_{\text{PD}}$.

Fig. 3.9 shows a schematic of such a plot. Drawing a horizontal line through the intercept demarcates film diffusion from pore diffusion and reaction rate. Below that horizontal line, the catalytic process is limited by a combination of pore diffusion rate and reaction rate. We must know which resistance controls product formation before scaling the process.

Pilot plant fixed-bed reactors are traditionally designed at space velocities equivalent to commercial scale fixed-bed reactors. Thus the film diffusion resistance of the process at the two scales is different. In general, pilot plant fixed-bed reactors are film diffusion rate limited, whereas commercial-size fixed-bed reactors are either pore diffusion rate or, more rarely, reaction rate limited. This shift from film diffusion rate limited to, more generally, pore diffusion rate limited occurs due to the high volumetric fluid flow through the catalyst mass in a commercial-size fixed-bed reactor. Thus reactant consumption or product formation is faster in the commercial-size fixed-bed reactor than in the pilot plant fixed-bed reactor.

The same shift occurs when downsizing a process from the commercial scale to the pilot plant scale. If we use a space velocity equivalent

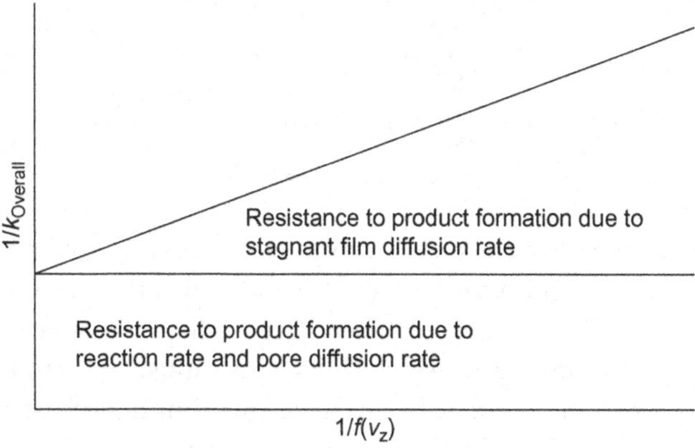

Figure 3.9 $1/k_{Overall}$ *as a function of* $1/f(v_z)$.

to that of the commercial fixed-bed reactor to design a pilot plant, then the controlling resistance to reactant consumption or product formation shifts from pore diffusion or reaction rate to stagnant film diffusion rate. This shift adversely impacts the results of the research program for which the pilot plant was built. Downsized pilot plants are built to solve process problems or to develop new catalysts. If the process problems are related to stagnant film diffusion, then such a pilot plant will be useful; however, if the process problems are not related to stagnant film diffusion, then the pilot plant will produce spurious and misleading information.

For the reasons stated earlier, it is important that we determine which regime controls the operating behavior of our model fixed-bed reactor. If we do not know which regime controls our model process, then our scaled prototype will most likely not behave similarly to it. Such an outcome can necessitate expensive modifications to our newly commissioned prototype, an outcome guaranteed to displease corporate management.

SUMMARY

Successfully scaling a process depends upon establishing the equality of the dimensionless parameters that are relevant to it. Those dimensionless parameters are

$$\begin{aligned}
\Pi_{\mathrm{M}}^{\mathrm{Geometric}} &= \Pi_{\mathrm{P}}^{\mathrm{Geometric}} \\
\Pi_{\mathrm{M}}^{\mathrm{Static}} &= \Pi_{\mathrm{P}}^{\mathrm{Static}} \\
\Pi_{\mathrm{M}}^{\mathrm{Kinematic}} &= \Pi_{\mathrm{P}}^{\mathrm{Kinematic}} \\
\Pi_{\mathrm{M}}^{\mathrm{Dynamic}} &= \Pi_{\mathrm{P}}^{\mathrm{Dynamic}} \\
\Pi_{\mathrm{M}}^{\mathrm{Thermal}} &= \Pi_{\mathrm{P}}^{\mathrm{Thermal}} \\
\Pi_{\mathrm{M}}^{\mathrm{Chemical}} &= \Pi_{\mathrm{P}}^{\mathrm{Chemical}}
\end{aligned} \tag{3.100}$$

Many of these dimensionless parameters contain overall rate constants, which are composite terms. These overall rate constants contain information about each step of the process. Thus before using them to calculate a given dimensionless parameter, we must determine which of those terms limits the process. We call the step limiting of the process the controlling regime. The controlling regime for the model and the prototype must be the same to ensure that the two processes behave similarly.

REFERENCES

[1] R. Clift, J. Grace, M. Weber, Bubbles, Drops, and Particles, Dover Publications Inc, Mineola, NY, 2005 (Chapter 1; originally published by Academic Press, Inc., 1978).

[2] G. Tatterson, Fluid Mixing and Gas Dispersion in Agitated Tanks, Second Printing, Greensboro, NC, 1991 (Chapters 1 and 2).

[3] R. Granger, Fluid Mechanics, Dover Publications Inc, Mineola, NY, 1995, p. 708 (Originally published by Holt, Reinhart and Winston, 1985).

[4] J. Miller, Separation Methods in Chemical Analysis, John Wiley & Sons, Inc, New York, NY, 1975, p. 6.

[5] R. Treybal, Mass Transfer Operations, McGraw-Hill Book Company, New York, NY, 1955, pp. 383–391.

[6] J. Carberry, First order rate processes and axial dispersion in packed bed reactors, Can. J. Chem. Eng. 36 (1958) 207.

[7] J. Carberry, Chemical and Catalytic Reaction Engineering, McGraw-Hill Book Company, New York, NY, 1976, pp. 530–531.

[8] R. Aris, The Mathematical Theory of Diffusion and Reaction in Permeable Catalysts, vol. 1, Clarendon Press, Oxford, UK, 1975.

[9] J. Knudsen, D. Katz, Fluid Dynamics and Heat Transfer, McGraw-Hill Book Company, New York, NY, 1958, pp. 257–258.

Scaling Fluid Flow

INTRODUCTION

Chemical engineering is about delivering feed to a chemical processing unit operation, converting that feed to product, then delivering product to customers. In the commodity chemical industry, nearly all feed deliveries involve a fluid, either a gas or a liquid. Similarly, delivery of commodity chemical products generally involves fluid movement. The exception is polymer products, which are generally delivered as small, solid pellets. Thus our interest is in fluid flow.

GENERAL FLUID FLOW CHARACTERISTICS

Most time-independent fluid flow depends upon pressure difference, ΔP [$ML^{-1}T^{-2}$]; gravitational acceleration, g [LT^{-2}]; fluid density, ρ [ML^{-3}]; viscosity, μ [$L^{-1}MT^{-1}$]; surface tension, σ [MT^{-2}]; compressibility as bulk modulus, β [$ML^{-1}T^{-2}$]; linear size, L [L]; and fluid velocity, v [LT^{-1}]. We will use dimensional analysis to determine the most common engineering descriptions for fluid flow.

For the above variables, the Dimension Table is

Variables		ΔP	g	L	σ	β	v	ρ	μ
Dimensions	L	−1	1	1	0	−1	1	−3	−1
	M	1	0	0	1	1	0	1	1
	T	−2	−2	0	−2	−2	−1	0	−1

and the Dimension matrix is

$$\begin{bmatrix} -1 & 1 & 1 & 0 & -1 & 1 & -3 & -1 \\ 1 & 0 & 0 & 1 & 1 & 0 & 1 & 1 \\ -2 & -2 & 0 & -2 & -2 & -1 & 0 & -1 \end{bmatrix} \tag{4.1}$$

The Rank matrix is the largest, nonsingular, square matrix contained in the Dimension matrix. Since the Dimension has three rows,

© 2016 Elsevier Inc. All rights reserved.

the largest square matrix will be a 3×3 matrix. Starting with the rightmost column of the Dimension matrix, the Rank matrix is

$$R = \begin{bmatrix} 1 & -3 & -1 \\ 0 & 1 & 1 \\ -1 & 0 & -1 \end{bmatrix} \tag{4.2}$$

and its determinant, calculated by a free-for-use matrix calculator on the Internet, is

$$|R| = \begin{vmatrix} 1 & -3 & -1 \\ 0 & 1 & 1 \\ -1 & 0 & -1 \end{vmatrix} = 3 \tag{4.3}$$

Thus the determinant is nonsingular and the number of dimensionless parameters for the general characteristics of fluid flow is

$$N_{\mathrm{p}} = N_{\mathrm{Var}} - R = 8 - 3 = 5 \tag{4.4}$$

The inverse of the Rank matrix is

$$R^{-1} = \begin{bmatrix} 1 & -3 & -1 \\ 0 & 1 & 1 \\ -1 & 0 & -1 \end{bmatrix}^{-1} = \begin{bmatrix} -1 & -3 & -2 \\ -1 & -2 & -1 \\ 1 & 3 & 1 \end{bmatrix} \tag{4.5}$$

and the Bulk matrix is

$$B = \begin{bmatrix} -1 & 1 & 1 & 0 & -1 \\ 1 & 0 & 0 & 1 & 1 \\ -2 & -2 & 0 & -2 & -2 \end{bmatrix} \tag{4.6}$$

Multiplying $-R^{-1}$ and B yields

$$\begin{aligned} -R^{-1} \cdot B &= -\begin{bmatrix} -1 & -3 & -2 \\ -1 & -2 & -1 \\ 1 & 3 & 1 \end{bmatrix} \begin{bmatrix} -1 & 1 & 1 & 0 & -1 \\ 1 & 0 & 0 & 1 & 1 \\ -2 & -2 & 0 & -2 & -2 \end{bmatrix} \\ &= \begin{bmatrix} -2 & -3 & 1 & -1 & -2 \\ -1 & -1 & 1 & 0 & -1 \\ 0 & 1 & -1 & -1 & 0 \end{bmatrix} \end{aligned} \tag{4.7}$$

The Total matrix is

$$T = \begin{bmatrix} I & 0 \\ -R^{-1} \cdot B & R^{-1} \end{bmatrix} \tag{4.8}$$

where I is the Identity matrix; 0 is the Zero matrix; and the matrices $-R^{-1} \cdot B$ and R^{-1} are Eq. (4.7) and Eq. (4.5), respectively. Completing the Total matrix give us

$$\begin{array}{c} \\ \Delta P \\ g \\ L \\ \sigma \\ \beta \\ v \\ \rho \\ \mu \end{array} \begin{array}{c} \begin{array}{ccccccccc} \Pi_1 & \Pi_2 & \Pi_3 & \Pi_4 & \Pi_5 & & & \end{array} \\ \begin{bmatrix} 1 & 0 & 0 & 0 & 0 & 0 & 0 & 0 \\ 0 & 1 & 0 & 0 & 0 & 0 & 0 & 0 \\ 0 & 0 & 1 & 0 & 0 & 0 & 0 & 0 \\ 0 & 0 & 0 & 1 & 0 & 0 & 0 & 0 \\ 0 & 0 & 0 & 0 & 1 & 0 & 0 & 0 \\ -2 & -3 & 1 & -1 & -2 & 1 & 3 & 2 \\ -1 & -1 & 1 & 0 & -1 & -1 & 2 & 1 \\ 0 & 1 & -1 & -1 & 0 & -1 & -3 & -1 \end{bmatrix} \end{array} \tag{4.9}$$

The dimensionless parameters are identified above each of the columns of the Identity matrix. The dimensionless parameters are

$$\Pi_1 = \frac{\Delta P}{v^2 \rho} \qquad \Pi_2 = \frac{g\mu}{v^3 \rho} \qquad \Pi_3 = \frac{\rho L v}{\mu} \qquad \Pi_4 = \frac{\sigma}{\mu v} \qquad \Pi_5 = \frac{\beta}{v^2 \rho}$$

Are these five dimensionless parameters independent of each other? The test for linear independence is

$$\alpha \Pi_1 + \beta \Pi_2 + \gamma \Pi_3 + \delta \Pi_4 + \varepsilon \Pi_5 = 0 \tag{4.10}$$

Substituting the matrix Π_1 through Π_5 columns into Eq. (4.10) gives

$$\alpha \begin{bmatrix} 1 \\ 0 \\ 0 \\ 0 \\ 0 \\ -2 \\ -1 \\ 0 \end{bmatrix} + \beta \begin{bmatrix} 0 \\ 1 \\ 0 \\ 0 \\ 0 \\ -3 \\ -1 \\ 1 \end{bmatrix} + \gamma \begin{bmatrix} 0 \\ 0 \\ 1 \\ 0 \\ 0 \\ 1 \\ 1 \\ -1 \end{bmatrix} + \delta \begin{bmatrix} 0 \\ 0 \\ 0 \\ 1 \\ 0 \\ -1 \\ 0 \\ -1 \end{bmatrix} + \varepsilon \begin{bmatrix} 0 \\ 0 \\ 0 \\ 0 \\ 1 \\ -2 \\ -1 \\ 0 \end{bmatrix} = \begin{bmatrix} 0 \\ 0 \\ 0 \\ 0 \\ 0 \\ 0 \\ 0 \\ 0 \end{bmatrix} \tag{4.11}$$

For Π_1, Π_2, Π_3, Π_4, and Π_5 to be independent, α, β, γ, δ, and ε must all be zero. Solving Eq. (4.11) yields

$$
\begin{aligned}
&\alpha + 0 + 0 + 0 + 0 = 0 \\
&0 + \beta + 0 + 0 + 0 = 0 \\
&0 + 0 + \gamma + 0 + 0 = 0 \\
&0 + 0 + 0 + \delta + 0 = 0 \\
&0 + 0 + 0 + 0 + \varepsilon = 0 \\
&-2\alpha - 3\beta + \gamma - \delta - 2\varepsilon = 0 \\
&-\alpha - \beta + \gamma - \varepsilon = 0 \\
&\beta - \gamma - \delta = 0
\end{aligned}
\tag{4.12}
$$

The above system of linear equations shows that $\alpha = \beta = \varepsilon = \delta = \varepsilon = 0$; thus all the dimensionless parameters are independent of each other. This result is important since it allows us to multiply and divide the individual dimensionless parameters for a given set to generate recognizable parameters. For example, Π_1 is the Euler number, Π_3 is the Reynolds number, and Π_5 is the inverse of the Cauchy number. However, Π_2 and Π_4 are not recognized, named dimensionless numbers, but because this group of dimensionless parameters forms a basis set, that is, independent of each other, we can multiply and divide them to obtain new dimensionless parameters. Our only constraint is: our analysis must produce five dimensionless parameters. For example, multiplying Π_2 by Π_3 gives us

$$
\Pi_2 \cdot \Pi_3 = \Pi_{2\cdot 3} = \left(\frac{g\mu}{v^3\rho}\right) \cdot \left(\frac{\rho L v}{\mu}\right) = \frac{gL}{v^2} \tag{4.13}
$$

an inverted form of the Froude number. Another form of the Froude number is $v/\sqrt{gL}$. Dividing Π_4 by Π_3 provides

$$
\frac{\Pi_4}{\Pi_3} = \Pi_{4/3} = \frac{\sigma/\mu v}{\rho L v/\mu} = \frac{\sigma}{\rho L v^2} \tag{4.14}
$$

which is the inverse of the Weber number.

Process scaling of flow characteristics involves some or all of these dimensionless parameters. In function notation, the description of general fluid flow is

$$
f(\Pi_1, \Pi_{2\cdot 3}, \Pi_3, \Pi_{4/3}, \Pi_5) = 0 \tag{4.15}
$$

One or all of these dimensionless parameters must be considered when scaling fluid flow. However, it is unlikely that all these dimensionless parameters will occur in a given scaling effort. If fluid compressibility is not a factor in a process, then the Cauchy number will not be pertinent to its process scaling. Also, if a process does not depend upon surface tension, then scaling the process will be independent of the Weber number. When a particular situation is specified, then scaling also requires geometric dimensionless parameters in order to meet the criteria for geometric similarity.

SCALING WATER FLOW IN SMOOTH PIPE

Let us assume you are the mayor of the small suburban town in which you live. It is an unpaid position, but you are a conscientious citizen fulfilling your social obligation toward your fellow citizens. Your municipal water comes from a local water reservoir created by damming a nearby river. The current supply pipeline from the reservoir to your town was installed during the late 1960s. Town growth has been such that water demand now exceeds water supply capacity. Your town needs a new water supply pipeline. Prior to contracting a civil engineering firm to design and install the new water supply pipeline, you decide to analyze the project first.

The current supply pipeline is smooth PVC pipe. You plan to replace it with larger diameter, smooth PVC pipe. The future supply pipeline will be installed parallel to the current supply pipeline; therefore, their lengths will be equivalent. You state your problem mathematically as

$$f(D, Q, \Delta P/L, \mu, \rho) = 0 \tag{4.16}$$

where D is pipe diameter with dimension [L]; Q is volumetric flow rate $[L^3T^{-1}]$; $\Delta P/L$ is pressure difference per unit length of pipe $[ML^{-2}T^{-2}]$; μ is fluid viscosity $[ML^{-1}T^{-1}]$; and ρ is fluid density $[ML^{-3}]$.

The Dimension Table for your analysis is

Variables		D	Q	$\Delta P/L$	μ	ρ
Dimensions	L	1	3	−2	−1	−3
	M	0	0	1	1	1
	T	0	−1	−2	−1	0

Your Dimension matrix is

$$D = \begin{bmatrix} 1 & 3 & -2 & -1 & -3 \\ 0 & 0 & 1 & 1 & 1 \\ 0 & -1 & -2 & -1 & 0 \end{bmatrix} \tag{4.17}$$

As with all Dimension matrices, the Rank matrix comprises those columns furthest right that form a square matrix. The Rank matrix for the above Dimension matrix is

$$R = \begin{bmatrix} -2 & -1 & -3 \\ 1 & 1 & 1 \\ -2 & -1 & 0 \end{bmatrix} \tag{4.18}$$

The Bulk matrix constitutes the remaining columns of a Dimension matrix, that is,

$$B = \begin{bmatrix} 1 & 3 \\ 0 & 0 \\ 0 & -1 \end{bmatrix} \tag{4.19}$$

If the determinant of R is nonsingular, that is, not zero, then the rank of the Dimension matrix is equal to the rows or columns of the R matrix. The determinant of R, Eq. (4.18), is

$$R = \begin{vmatrix} -2 & -1 & -3 \\ 1 & 1 & 1 \\ -2 & -1 & 0 \end{vmatrix} = -3 \tag{4.20}$$

Thus the above Dimension matrix has a rank of 3. The mathematical statement for your analysis contains five variables; therefore, this Dimensional Analysis will generate two dimensionless parameters, per Buckingham's Pi Theorem, which is

$$\begin{aligned} N_P &= N_{Var} - R \\ N_P &= 5 - 3 = 2 \end{aligned} \tag{4.21}$$

The inverse of the above Rank matrix is

$$R^{-1} = \begin{bmatrix} -0.33 & -1 & -0.66 \\ 0.66 & 2 & 0.33 \\ -0.33 & 0 & 0.33 \end{bmatrix} \tag{4.22}$$

Matrix multiplying $-R^{-1}$ and B gives you

$$-R^{-1}\cdot B = -\begin{bmatrix} -0.33 & -1 & -0.66 \\ 0.66 & 2 & 0.33 \\ -0.33 & 0 & 0.33 \end{bmatrix} \cdot \begin{bmatrix} 1 & 3 \\ 0 & 0 \\ 0 & -1 \end{bmatrix} = \begin{bmatrix} 0.33 & 0.33 \\ -0.66 & -1.66 \\ 0.33 & 1.33 \end{bmatrix} \tag{4.23}$$

Your Total matrix is

$$T = \begin{bmatrix} I & 0 \\ -R^{-1}\cdot B & R^{-1} \end{bmatrix} \tag{4.24}$$

where I is the Identity matrix; 0 is the Zero matrix; and the matrices $-R^{-1}\cdot B$ and R^{-1} come from your analysis. Filling your Total matrix gives you

$$T = \begin{matrix} & \Pi_1 & \Pi_2 & & & \\ D & 1 & 0 & 0 & 0 & 0 \\ Q & 0 & 1 & 0 & 0 & 0 \\ \Delta P/L & 0.33 & 0.33 & -0.33 & -1 & -0.66 \\ \mu & -0.66 & -1.66 & 0.66 & 2 & 0.33 \\ \rho & 0.33 & 1.33 & -0.33 & 0 & 0.33 \end{matrix} \tag{4.25}$$

Your dimensionless parameters are the columns comprising the Identity matrix in the upper left portion of the Total matrix. They are identified as Π_1 and Π_2 in Eq. (4.25). The variables associated with each row are noted to the left of the Total matrix. The above format makes it easy to formulate the dimensionless parameters, they are

$$\Pi_1 = \frac{D(\Delta P/L)^{0.33}\rho^{0.33}}{\mu^{0.66}} \quad \text{and} \quad \Pi_1 = \frac{Q(\Delta P/L)^{0.33}\rho^{1.33}}{\mu^{1.66}} \tag{4.26}$$

Your preference is to avoid dimensionless parameters with variables raised to fractional powers. Since the Total matrix yields dimensionless parameters that are independent of each other, you can combine them to form a new set of independent dimensionless parameters. Your only constraint is that you generate N_P of them.

Manipulating the two dimensionless parameters gives you

$$\Pi_1^3 = \frac{D^3(\Delta P/L)\rho}{\mu^2} \quad \text{and} \quad \frac{\Pi_2}{\Pi_1} = \Pi_{2/1} = \frac{Q\rho}{D\mu} \tag{4.27}$$

To determine the dimensions of your future water supply pipeline, you set

$$
\begin{aligned}
(\Pi_1^3)_{\text{Current}} &= (\Pi_1^3)_{\text{Future}} \\
(\Pi_{2/1})_{\text{Current}} &= (\Pi_{2/1})_{\text{Future}}
\end{aligned}
\tag{4.28}
$$

You know Q_{Current} and you will specify Q_{Future}; therefore

$$
\frac{Q_C \rho_C}{D_C \mu_C} = \frac{Q_F \rho_F}{D_F \mu_F} \tag{4.29}
$$

where the subscript C indicates the current pipeline and the subscript F identifies the future pipeline. Note that, for this analysis $\rho_C = \rho_F$ and $\mu_C = \mu_F$. Therefore, Eq. (4.29) becomes

$$
\frac{Q_C}{D_C} = \frac{Q_F}{D_F} \tag{4.30}
$$

And, by rearranging Eq. (4.30), you determine the diameter of your future supply pipeline for a specified Q_F; the result is

$$
D_F = \frac{Q_F D_C}{Q_C} \tag{4.31}
$$

From Π_1^3, you can determine the pressure difference per unit length of pipe for the future pipeline, which is

$$
\begin{aligned}
(\Pi_1^3)_{\text{Current}} &= (\Pi_1^3)_{\text{Future}} \\
\frac{D_C^3 (\Delta P/L)_C \rho_C}{\mu_C^2} &= \frac{D_F^3 (\Delta P/L)_F \rho_F}{\mu_F^2}
\end{aligned}
\tag{4.32}
$$

Eq. (4.32) reduces to

$$
D_C^3 (\Delta P/L)_C = D_F^3 (\Delta P/L)_F \tag{4.33}
$$

Rearranging Eq. (4.33) gives you

$$
(\Delta P/L)_F = \frac{D_C^3 (\Delta P/L)_C}{D_F^3} \tag{4.34}
$$

You know the values for your current variables and you can calculate D_F from Eq. (4.31). With this information, you are now ready to issue bids for your water supply project.

SCALING FLUID FLOW IN A CIRCULAR PIPE

All chemical engineers at some point in their careers become interested in fluid flow in circular pipes. The variable of most concern is pressure difference between a starting location and a final location. In petrochemical plants and along transport pipelines, there is only so much pressure available to induce flow. In other words, pressure difference; i.e., pressure loss must be minimized. Thus the chemical engineer's interest is in pressure difference along any segment of pipe.

The geometric parameters for pipe flow are pipe diameter D [L], pipe length L [L], and pipe wall roughness r [L]. The material parameters are fluid density ρ [$L^{-3}M$] and fluid viscosity μ [$L^{-1}MT^{-1}$]. The process parameters are fluid velocity v [LT^{-1}] and pressure difference per pipe length $\Delta P/L$ [$L^{-2}MT^{-2}$].

The Dimension Table is

Variables		$\Delta P/L$	D	L	r	v	ρ	μ
Dimensions	L	−2	1	1	1	1	−3	−1
	M	1	0	0	0	0	1	1
	T	−2	0	0	0	−1	0	−1

and the Dimension matrix is

$$\begin{bmatrix} -2 & 1 & 1 & 1 & 1 & -3 & -1 \\ 1 & 0 & 0 & 0 & 0 & 1 & 1 \\ -2 & 0 & 0 & 0 & -1 & 0 & -1 \end{bmatrix} \tag{4.35}$$

The Rank matrix is the largest, nonsingular, square matrix contained in the Dimension matrix. Since the Dimension has three rows, the largest square matrix will be a 3×3 matrix. Starting with the rightmost column of the Dimension matrix, the Rank matrix is

$$R = \begin{bmatrix} 1 & -3 & -1 \\ 0 & 1 & 1 \\ -1 & 0 & -1 \end{bmatrix} \tag{4.36}$$

and its determinant, calculated by a free-for-use matrix calculator on the Internet, is

$$|R| = \begin{vmatrix} 1 & -3 & -1 \\ 0 & 1 & 1 \\ -1 & 0 & -1 \end{vmatrix} = 3 \tag{4.37}$$

Thus the determinant is nonsingular and the number of dimensionless parameters is

$$N_P = N_{Var} - R = 7 - 3 = 4 \tag{4.38}$$

The inverse of the Rank matrix is

$$R^{-1} = \begin{bmatrix} 1 & -3 & -1 \\ 0 & 1 & 1 \\ -1 & 0 & -1 \end{bmatrix}^{-1} = \begin{bmatrix} -1 & -3 & -2 \\ -1 & -2 & -1 \\ 1 & 3 & 1 \end{bmatrix} \tag{4.39}$$

and the Bulk matrix is

$$B = \begin{bmatrix} -2 & 1 & 1 & 1 \\ 1 & 0 & 0 & 0 \\ -2 & 0 & 0 & 0 \end{bmatrix} \tag{4.40}$$

Multiplying $-R^{-1}$ and B gives us

$$-R^{-1} \cdot B = -\begin{bmatrix} -1 & -3 & -2 \\ -1 & -2 & -1 \\ 1 & 3 & 1 \end{bmatrix} \cdot \begin{bmatrix} -2 & 1 & 1 & 1 \\ 1 & 0 & 0 & 0 \\ -2 & 0 & 0 & 0 \end{bmatrix} = \begin{bmatrix} -3 & 1 & 1 & 1 \\ -2 & 1 & 1 & 1 \\ 1 & -1 & -1 & -1 \end{bmatrix} \tag{4.41}$$

The Total matrix is, then

$$\begin{array}{c} \\ \Delta P/L \\ D \\ L \\ r \\ v \\ \rho \\ \mu \end{array} \begin{array}{c} \begin{array}{cccc} \Pi_1 & \Pi_2 & \Pi_3 & \Pi_4 \end{array} \\ \begin{bmatrix} 1 & 0 & 0 & 0 & 0 & 0 & 0 \\ 0 & 1 & 0 & 0 & 0 & 0 & 0 \\ 0 & 0 & 1 & 0 & 0 & 0 & 0 \\ 0 & 0 & 0 & 1 & 0 & 0 & 0 \\ -3 & 1 & 1 & 1 & -1 & -3 & -2 \\ -2 & 1 & 1 & 1 & -1 & -2 & -1 \\ 1 & -1 & -1 & -1 & 1 & 3 & 1 \end{bmatrix} \end{array} \tag{4.42}$$

This example has four dimensionless parameters; therefore, the Identity matrix has four columns, which are denoted by Π_1, Π_2, Π_3, and Π_4. Matching variables with column elements in the Total matrix gives us the dimensionless parameters, which are

$$\Pi_1 = \frac{(\Delta P/L)\mu}{v^3\rho^2} \qquad \Pi_2 = \frac{\rho D v}{\mu} \qquad \Pi_3 = \frac{L v \rho}{\mu} \qquad \Pi_4 = \frac{r v \rho}{\mu}$$

Note that Π_2 is the Reynolds number.

Since these dimensionless parameters are independent of each other, we can combine them to clarify the physics of the result. Doing so yields three new dimensionless parameters, which are

$$\frac{\Pi_3}{\Pi_2} = \Pi_{3/2} = \frac{L\, v\, \rho/\mu}{D\, v\, \rho/\mu} = \frac{L}{D}$$
$$\frac{\Pi_4}{\Pi_2} = \Pi_{4/2} = \frac{r\, v\, \rho/\mu}{D\, v\, \rho/\mu} = \frac{r}{D} \qquad (4.43)$$
$$\Pi_1\Pi_2 = \Pi_{12} = \left(\frac{(\Delta P/L)\mu}{v^3\rho^2}\right)\left(\frac{D\, v\, \rho}{\mu}\right) = \frac{D(\Delta P/L)}{\rho v^2} = 2f$$

where $2f$ if the Fanning friction number. The solution for this example, in function notation, is

$$f(\Pi_{12}, \Pi_{3/2}, \Pi_{4/2}, \Pi_2) = f\left(\frac{D(\Delta P/L)}{\rho v^2}, \frac{L}{D}, \frac{r}{D}, \frac{\rho\, D\, v}{\mu}\right) = 0 \qquad (4.44)$$

Writing the function in terms of Π_{12} gives us

$$\Pi_{12} = 2f = \kappa \cdot f(\Pi_{3/2}, \Pi_{4/2}, \Pi_2) = \kappa \cdot f\left(\frac{L}{D}, \frac{r}{D}, \frac{\rho\, D\, v}{\mu}\right) \qquad (4.45)$$

or

$$\frac{D(\Delta P/L)}{\rho v^2} = 2f = \kappa \cdot f(\Pi_{3/2}, \Pi_{4/2}, \Pi_2) \qquad (4.46)$$

where κ and $f(\Pi_{3/2}, \Pi_{4/2}, \Pi_2)$ are determined by experimentation. The most common plots are $2f$ versus Π_2, that is, the Reynolds number, with r/D, that is, $\Pi_{4/2}$, as the parametric line. A second plot would be $2f$ versus Π_2 with $\Pi_{3/2}$, that is, L/D, as the parametric line. Such plots are common in the engineering literature [1–3].

The above set of dimensionless parameters allows us to scale piping from a model to a prototype. Scaling pipe requires

$$(\Pi_{3/2})_M = (\Pi_{3/2})_P$$
$$(\Pi_{4/2})_M = (\Pi_{4/2})_P \qquad (4.47)$$
$$(\Pi_2)_M = (\Pi_2)_P$$
$$(\Pi_{12})_M = (\Pi_{12})_P$$

or, in terms of the variables, these requirements become

$$\begin{aligned}
\frac{L_M}{D_M} &= \frac{L_P}{D_P} \\
\frac{r_M}{D_M} &= \frac{r_P}{D_P} \\
\frac{\rho_M D_M v_M}{\mu_M} &= \frac{\rho_P D_P v_P}{\mu_P} \\
\frac{D_M(\Delta P/L)_M}{\rho_M v_M^2} &= \frac{D_P(\Delta P/L)_P}{\rho_P v_P^2}
\end{aligned} \tag{4.48}$$

where subscript M denotes model and subscript P denotes prototype.

We determine each variable by starting with the dimensionless parameter with the fewest variables. Thus we start with

$$\frac{L_M}{D_M} = \frac{L_P}{D_P} \tag{4.49}$$

because we know L_M and D_M from our model and we specify L_P for our prototype. Therefore, the diameter of the pipe for the prototype is

$$D_P = \left(\frac{L_P}{L_M}\right) D_M \tag{4.50}$$

We next determine the allowable wall roughness for the prototype. From above

$$\frac{r_M}{D_M} = \frac{r_P}{D_P} \tag{4.51}$$

Therefore, prototype wall roughness is

$$r_P = \left(\frac{D_P}{D_M}\right) r_M \tag{4.52}$$

Maintaining geometric similarity between the model and the prototype means wall roughness in the model must be proportionately less than in the prototype. In other words, we must use smoother pipe in the model. The penalty, measured by the ratio of friction factor for commercial pipe and smooth glass tube, for not maintaining geometric similarity with regard to wall roughness is negligible for laminar flow, that is, for low Reynolds numbers. The friction factor ratio is 1 for laminar flow [4]. However, the penalty increases with increasing Reynolds number. At 10^6 Reynolds number, the friction factor ratio is 1.65, indicating significant error in the calculated flow for the prototype [4]. Thus the difficulty with scaling fluid flow through circular pipes is maintaining geometric similarity with regard to wall roughness.

To determine the fluid velocity in the prototype, we use the equality

$$\frac{\rho_M D_M v_M}{\mu_M} = \frac{\rho_P D_P v_P}{\mu_P} \tag{4.53}$$

However, in most cases, $\rho_M = \rho_P$ and $\mu_M = \mu_P$. Therefore, Eq. (4.53) reduces to

$$D_M v_M = D_P v_P \tag{4.54}$$

which rearranges to

$$v_P = \left(\frac{D_M}{D_P}\right) v_M \tag{4.55}$$

We know D_M and v_M from our model and we calculated D_P from Eq. (4.50).

In some cases, when the model and prototype use the same fluid, it may be difficult to meet the Reynolds number criterion. For gas processes, the corresponding velocities in the model can induce compression effects in which case the Cauchy number becomes a scaling parameter. For liquids, where compression is negligible, maintaining high fluid velocity requires excessive power. In this case, using a fluid of lower kinematic viscosity in the model, while meeting the Reynolds number criterion, provides scalable information [4].

We are now able to calculate $(\Delta P/L)_P$ for our prototype. We start with

$$\frac{D_M(\Delta P/L)_M}{\rho_M v_M^2} = \frac{D_P(\Delta P/L)_P}{\rho_P v_P^2} \tag{4.56}$$

In most cases, $\rho_M = \rho_P$; therefore

$$\frac{D_M(\Delta P/L)_M}{v_M^2} = \frac{D_P(\Delta P/L)_P}{v_P^2} \tag{4.57}$$

which rearranges to

$$(\Delta P/L)_P = \left(\frac{v_P^2}{v_M^2}\frac{D_M}{D_P}\right)(\Delta P/L)_M \tag{4.58}$$

for which values are known for all the variables.

Plotting friction factor versus Reynolds number produces a figure with three flow regimes: a laminar flow region where friction factor relates linearly to Reynolds number; a transition flow region where friction factor relates nonlinearly to Reynolds number; and, a turbulent flow region, generally at $Re > 10{,}000$, where friction factor is independent of Reynolds number. It is relatively simple to meet the Reynolds number criterion for model and prototype in the laminar flow region. In this region, experimental data can be easily obtained from the model and reliably scaled into the prototype. However, for the turbulent flow region, it can be difficult to achieve model Reynolds numbers equivalent to prototype Reynolds numbers. Fortunately, since the friction factor is independent of Reynolds number in the turbulent flow region, mismatched Reynolds numbers do not cause an issue with regard to scaling. We only need to ensure that the model operates in the turbulent flow region in order to obtain model data scalable to the prototype. With regard to the transition flow region of a friction factor versus Reynolds number plot—avoid it. In the transition flow region, small differences in Reynolds number can cause large differences in friction factor, which induces problems in the prototype.

SCALING CENTRIFUGAL PUMPS

All chemical engineers use pumps, many of which utilize centrifugal force to increase fluid pressure. In fact, chemical engineers favor centrifugal pumps because of their efficiency and ruggedness. At some point in every chemical engineer's career, he or she will be asked to scale a centrifugal pump from a pilot plant or a semi-works to a commercial plant.

The geometric variable for a centrifugal pump is a characteristic length L, which is generally the diameter of the pump chamber; its dimension is [L]. The material variables are fluid viscosity, μ [$ML^{-1}T^{-1}$], and fluid density, ρ [ML^{-3}]. The process variables are revolutions per unit time, n [T^{-1}]; volumetric flow rate, Q [L^3T^{-1}]; pressure difference, ΔP [$ML^{-1}T^{-2}$]; and power input to the pump motor, W [ML^2T^{-3}].

The Dimension Table for scaling a centrifugal pump is

Variables		n	L	Q	ρ	μ	ΔP	W
Dimensions	L	0	1	3	−3	−1	−1	2
	M	0	0	0	1	1	1	1
	T	−1	0	−1	0	−1	−2	−3

Thus, the Dimension matrix is

$$\begin{bmatrix} 0 & 1 & 3 & -3 & -1 & -1 & 2 \\ 0 & 0 & 0 & 1 & 1 & 1 & 1 \\ -1 & 0 & -1 & 0 & -1 & -2 & -3 \end{bmatrix} \tag{4.59}$$

and the Rank matrix is

$$R = \begin{bmatrix} -1 & -1 & 2 \\ 1 & 1 & 1 \\ -1 & -2 & -3 \end{bmatrix} \tag{4.60}$$

The determinant of R is

$$|R| = \begin{vmatrix} -1 & -1 & 2 \\ 1 & 1 & 1 \\ -1 & -2 & -3 \end{vmatrix} = -3 \tag{4.61}$$

Thus the rank of the above Dimension matrix is 3. From Buckingham's Pi Theorem

$$N_{\mathrm{P}} = N_{\mathrm{Var}} - R = 7 - 3 = 4 \tag{4.62}$$

where we will generate four dimensionless parameters from this analysis.

We can calculate the inverse of the Rank matrix using the matrix calculation in Excel or we can calculate it using software available for free use on the Internet. The inverse of matrix R is

$$R^{-1} = \begin{bmatrix} 0.33 & 2.33 & 1 \\ -0.66 & -1.66 & -1 \\ 0.33 & 0.33 & 0 \end{bmatrix} \tag{4.63}$$

The Bulk matrix is

$$B = \begin{bmatrix} 0 & 1 & 3 & -3 \\ 0 & 0 & 0 & 1 \\ -1 & 0 & -1 & 0 \end{bmatrix} \tag{4.64}$$

Multiplying $-R^{-1}$ by the Bulk matrix gives us

$$R^{-1}\cdot B = \begin{bmatrix} 0.33 & 2.33 & 1 \\ -0.66 & -1.66 & -1 \\ 0.33 & 0.33 & 0 \end{bmatrix} \cdot \begin{bmatrix} 0 & 1 & 3 & -3 \\ 0 & 0 & 0 & 1 \\ -1 & 0 & -1 & 0 \end{bmatrix}$$
$$= \begin{bmatrix} 1 & -0.33 & 0 & -1.33 \\ -1 & 0.66 & 1 & -0.33 \\ 0 & -0.33 & -1 & 0.66 \end{bmatrix} \quad (4.65)$$

We can now assemble the Total matrix, which is

$$T = \begin{array}{c} \\ n \\ L \\ Q \\ \rho \\ \mu \\ \Delta P \\ W \end{array} \begin{array}{c} \begin{array}{cccc} \Pi_1 & \Pi_2 & \Pi_3 & \Pi_4 \end{array} \\ \begin{bmatrix} 1 & 0 & 0 & 0 & 0 & 0 & 0 \\ 0 & 1 & 0 & 0 & 0 & 0 & 0 \\ 0 & 0 & 1 & 0 & 0 & 0 & 0 \\ 0 & 0 & 0 & 1 & 0 & 0 & 00.66 \\ 1 & -0.33 & 0 & -1.33 & 0.33 & 2.33 & 1 \\ -1 & 0.66 & 1 & -0.33 & -0.66 & -1.66 & -1 \\ 0 & -0.33 & -1 & 0.66 & 0 & -0.33 & -1 \end{bmatrix} \end{array} \quad (4.66)$$

The dimensionless parameters are identified along the top of the Total matrix and the variables of this scaling analysis are placed along the left bracket of the Total matrix. The individual dimensionless parameters are

$$\Pi_1 = \frac{n\,\mu}{\Delta P}\,\Pi_2 = \frac{L(\Delta P)^{0.66}}{\mu^{0.33}\,W^{0.33}}\,\Pi_3 = \frac{Q(\Delta P)}{P}\,\Pi_4 = \frac{\rho\,W^{0.66}}{\mu^{1.33}(\Delta P)^{0.33}}$$

Combining these dimensionless parameters to remove the fractional powers gives us a different set of scaling parameters. This new set of dimensionless parameters is

$$\Pi_1 = \frac{n\,\mu}{\Delta P}$$
$$\Pi_4^3 = \frac{\rho^3\,W^2}{\mu^4(\Delta P)}$$
$$\Pi_1\Pi_2^3 = \left(\frac{n\,\mu}{\Delta P}\right)\left(\frac{L^3(\Delta P)^2}{\mu\,W}\right) = \frac{n(\Delta P)L^3}{W} \quad (4.67)$$
$$\frac{\Pi_3}{\Pi_1\Pi_2^3} = \frac{Q(\Delta P)/W}{(n\,\mu/\Delta P)(L^3(\Delta P)^2/\mu\,W)} = \frac{Q}{n\,L^3}$$

The scaling criterion is $\Pi^{\text{Model}}_{\text{Mechanical}} = \Pi^{\text{Prototype}}_{\text{Mechanical}}$. To scale equipment or a process, we start with the dimensionless parameter with the fewest variables. In this case, we start with Π_1.

$$(\Pi_1)_M = (\Pi_1)_P \tag{4.68}$$

where subscript M indicates the model and subscript P identifies the prototype. Substituting the appropriate variables into Eq. (4.68), we obtain

$$\frac{n_M \mu_M}{(\Delta P)_M} = \frac{n_P \mu_P}{(\Delta P)_P} \tag{4.69}$$

But, most likely, $\mu_M = \mu_P$; thus Eq. (4.69) reduces to

$$\frac{n_M}{(\Delta P)_M} = \frac{n_P}{(\Delta P)_P} \tag{4.70}$$

We know n_M and $(\Delta P)_M$ from our model. The process design engineers specify $(\Delta P)_P$ for our prototype; therefore, we solve Eq. (4.70) for n_P, which is

$$n_P = \left[\frac{(\Delta P)_P}{(\Delta P)_M}\right] n_M \tag{4.71}$$

We next calculate the power required by the prototype centrifugal pump. Π_4^3 is the dimensionless number with the fewest unknown variables including power W. Therefore

$$\begin{aligned} (\Pi_4^3)_M &= (\Pi_4^3)_P \\ \frac{\rho_M^3 W_M^2}{\mu_M^4 (\Delta P)_M} &= \frac{\rho_P^3 W_P^2}{\mu_P^4 (\Delta P)_P} \end{aligned} \tag{4.72}$$

But, most likely, $\mu_M = \mu_P$ and $\rho_M = \rho_P$; therefore, Eq. (4.72) becomes

$$\frac{W_M^2}{(\Delta P)_M} = \frac{W_P^2}{(\Delta P)_P} \tag{4.73}$$

and, solving for W_P yields

$$W_P = \left[\frac{(\Delta P)_P}{(\Delta P)_M}\right]^{1/2} W_M \tag{4.74}$$

We know $(\Delta P)_M$ from our model and our process design engineers specified $(\Delta P)_P$. Thus we can calculate W_P.

We can now calculate the characteristic length L for our prototype centrifugal pump. We make that calculation using $\Pi_1\Pi_2^3$.

$$(\Pi_1\Pi_2^3)_M = (\Pi_1\Pi_2^3)_P$$
$$\frac{n_M(\Delta P)_M L_M^3}{W_M} = \frac{n_P(\Delta P)_P L_P^3}{W_P} \tag{4.75}$$

All the above variables are known or calculated, except for L_P. Rearranging Eq. (4.75) for L_P gives

$$L_P = \left[\frac{n_M(\Delta P)_M}{n_P(\Delta P)_P}\frac{W_P}{W_M}\right]^{1/3} L_M \tag{4.76}$$

The last unknown variable is volumetric flow rate. We calculate it from $\Pi_3/\Pi_1\Pi_2^3$.

$$\left(\frac{\Pi_3}{\Pi_1\Pi_2^3}\right)_M = \left(\frac{\Pi_3}{\Pi_1\Pi_2^3}\right)_P$$
$$\frac{Q_M}{n_M L_M^3} = \frac{Q_P}{n_P L_P^3} \tag{4.77}$$

We know n_M, Q_M, and L_M from our model and we have already calculated n_P and L_P above. Therefore, solving for Q_P we get

$$Q_P = \left(\frac{n_P L_P^3}{n_M L_M^3}\right) Q_M \tag{4.78}$$

We have now completely specified the centrifugal pump for our prototype. Its operating behavior will be the same as the operating behavior of the centrifugal pump in our model.

FLUID FLOW THROUGH A PACKED BED

Many unit operations in a chemical process involve a fluid flowing through a circular column filled with material particles. The material particles can be a porous solid or a structured packing. We use porous solids as adsorbents to remove process poisons from process feed streams or to remove contaminant from product prior to selling it. We also use porous solids as support for catalytic metals, which we use in fixed-bed reactors. We use structured packing in any process involving mass transfer across a physical interface, for example, across a gas–liquid interface or a liquid–liquid interface. The structured packing increases the interfacial surface area available for mass transfer. Absorption and extraction columns are also filled with structured packing. We also put structured packing in portions of distillation columns; distillation columns containing only structured packing are not uncommon in the chemical processing industry.

From a process design viewpoint, the major concern with a column of material particles is pressure difference, ΔP, along the length of the material bed. Scaling such a column is one of the more difficult challenges confronting chemical engineers. Unit operations downstream of the packed column may not operate properly if the prototype packed column has a larger ΔP than predicted by the scaling procedure. Such an outcome requires retrofitting the chemical process, which is always expensive. Either larger pumps or compressors must be installed in the chemical process or the column must be emptied and refilled with material particles of higher voidage. Unfortunately, this latter option suffers from the same shortcoming as the original design, that is, how to estimate the prototype ΔP from the model ΔP.

In 1952, Sabri Ergun published an equation that has since been used for scaling packed columns [5]. Its purpose was to explain the transition zone between laminar flow through a mass of material particles and turbulent flow through the same mass of material particles. When we plot a function of pressure drop against the Reynolds number for a flowing fluid, we obtain a negatively sloped line for fluid flowing at low Reynolds numbers and a horizontal line for fluid flowing at high Reynolds numbers. We describe the former flow regime as laminar flow and the latter flow regime as turbulent flow. The experimental data curves smoothly from the laminar regime to the turbulent regime. The region between the laminar and turbulent flow regimes is the "transition" flow regime. Theoretically, the two flow regimes should make a sharp, obtuse intersection. Ergun's goal was to explain the smooth transition from the laminar flow regime to the turbulent flow regime.

We must derive the Ergun equation to understand its strengths and weaknesses [6]. Consider a column filled with material particles and imagine the interstitial space between the individual particles as forming small pipes or tubes through which fluid flows. If the fluid flowing through such a pipe or tube is at low Reynolds number, then from the Hagan–Poiseuille equation, the pressure drop, ΔP, along the conduit is

$$\Delta P = \frac{8\mu v_{\mathrm{Inter}} L}{r^2} \tag{4.79}$$

where μ is fluid viscosity [$\mathrm{L^{-1}MT^{-1}}$]; v_{Inter} is the interstitial fluid velocity [$\mathrm{LT^{-1}}$], that is, the velocity of the fluid flowing between the material particles; L is conduit [L], that is, tube, length; and r is pipe or tube radius [L].

Pressure difference indicates flow efficiency. Inefficient flow has a large pressure difference while efficient flow has a small pressure drop. Flow efficiency depends on conduit or channel structure, which depends upon channel breadth and channel length. Unfortunately, we characterize flow by one channel variable, usually the radius or diameter of the flow area. We use "hydraulic radius" to better characterize a flow channel [7]. Hydraulic radius for a flow channel is defined as

$$r_{\mathrm{H}} = \frac{\text{Cross-sectional area open to flow}}{\text{Wetted perimeter}} \tag{4.80}$$

which is, for a circular conduit

$$r_{\mathrm{H}} = \frac{\pi r^2}{2\pi r} = \frac{r}{2} \tag{4.81}$$

Substituting into the Hagen–Poiseuille equation above gives us

$$\Delta P = \frac{8\mu v_{\mathrm{Inter}} L}{4 r_{\mathrm{H}}^2} \tag{4.82}$$

If we multiply r_{H} by 1, that is, by L/L, then r_{H} becomes the ratio of the void volume of the packing material to the total surface area of the packing material [8]. In other words, the hydraulic radius can now be defined in terms of the packing material [9]

$$r_{\mathrm{H}} = \frac{\text{Volume open to flow}}{(\text{Number of material particles}) \cdot (\text{Surface area of one particle})} \tag{4.83}$$

where

$$\text{Volume open to flow} = \varepsilon V_{\mathrm{B}} \tag{4.84}$$

and

$$\text{Number of material particles} = \frac{(1-\varepsilon)V_B}{\text{Volume of one material particle}} \tag{4.85}$$

In Eqs. (4.84) and (4.85) ε is void fraction and V_B is material particle bed volume. The equation for r_H now becomes

$$r_H = \frac{\varepsilon V_B}{\left(\dfrac{(1-\varepsilon)V_B}{\text{Volume of one material particle}}\right) \cdot (\text{Surface area of one particle})}$$

$$r_H = \frac{\varepsilon V_B}{(1-\varepsilon)V_B \cdot \left(\dfrac{\text{Surface area of one particle}}{\text{Volume of one material particle}}\right)} = \frac{\varepsilon}{(1-\varepsilon)(S/V)} \tag{4.86}$$

where S is the surface area and V is the volume of the material particle. The S/V ratio for spheres is

$$\frac{S}{V} = \frac{4\pi r^2}{(4/3)\pi r^3} = \frac{3}{r_S} \tag{4.87}$$

Therefore, the hydraulic radius for spherical particles becomes

$$r_H = \left(\frac{\varepsilon}{1-\varepsilon}\right)\frac{1}{(3/r_S)} = \left(\frac{\varepsilon}{1-\varepsilon}\right)\frac{r_S}{3} \tag{4.88}$$

Substituting hydraulic radius r_H for r in the Hagan–Poiseuille equation gives

$$\Delta P = \frac{8\mu v L}{4\left[(\varepsilon/(1-\varepsilon))(r_S/3)\right]^2} = \left(\frac{1-\varepsilon}{\varepsilon}\right)^2\left(\frac{72\mu v L}{4r_S^2}\right) \tag{4.89}$$

But, $4r_S^2 = d_S^2$, where d_S is the diameter of the spherical packing. Substituting d_S for r_S in Eq. (4.89) gives us

$$\Delta P = \frac{72\mu v L}{d_S^2}\frac{(1-\varepsilon)^2}{\varepsilon^2} \tag{4.90}$$

Converting fluid velocity through the catalyst mass to fluid velocity through the empty reactor yields

$$\Delta P = \frac{72\mu v_{\text{Super}} L}{d_S^2} \frac{(1-\varepsilon)^2}{\varepsilon^3} \tag{4.91}$$

where $v_{\text{Super}} = v_{\text{Inter}}/\varepsilon$. v_{Super} is the superficial fluid velocity.

In Eq. (4.91), we assumed a linear path through the mass of the material particle, that path being given by L. But, no flow path through the material mass is straight, every path involves many twists and turns. We say the flow path is "tortuous." Ergun and others estimate that tortuosity to be

$$L = (25/12)Z \tag{4.92}$$

where Z is the physical height of the material packing. Applying this adjustment to Eq. (4.91) yields

$$\Delta p = \frac{72\mu v_{\text{Super}}(25/12)Z}{d_S^2} \frac{(1-\varepsilon)^2}{\varepsilon^3} = \frac{150\mu v_{\text{Super}} Z}{d_S^2} \frac{(1-\varepsilon)^2}{\varepsilon^3} \tag{4.93}$$

Eq. (4.93) is the Blake–Kozeny equation, which is valid only for the laminar flow regime [10].

The pressure drop for turbulent flow through a channel, such as a duct, is given as

$$\Delta P = \frac{f \rho Z v^2}{2D} \tag{4.94}$$

where f is the friction factor for the flow; ρ is fluid density; Z is the physical height of the packed material (note that we are taking Z to be the length of the flow conduit); v is fluid velocity; and D is channel diameter. As before, the channels through a mass of material particles possess irregular shapes; therefore, we must use hydraulic radius as the characteristic length to describe them. Assuming the channels formed through the catalyst mass have circular shapes allows us to use the above defined formula for hydraulic radius. Substituting into Eq. (4.94) gives

$$\Delta P = \frac{f \rho Z v^2}{8 r_H} \tag{4.95}$$

But, for spherical material particles

$$r_{\mathrm{H}} = \left(\frac{\varepsilon}{1-\varepsilon}\right)\frac{d_{\mathrm{S}}}{6} \tag{4.96}$$

where d_{S} is the diameter of the spheres. Thus the pressure drop becomes

$$\Delta P = \frac{f\rho Z v^2}{(8/6)d_{\mathrm{S}}(\varepsilon/1-\varepsilon)} = \left(\frac{3f}{2}\right)\left(\frac{\rho Z v^2}{2d_{\mathrm{S}}}\right)\left(\frac{1-\varepsilon}{\varepsilon}\right) \tag{4.97}$$

Many studies concerning turbulent flow indicate that $3f/2$ is equivalent to 3.5 [6]. Making that substitution yields

$$\Delta P = 3.5\left(\frac{\rho Z v^2}{2d_{\mathrm{S}}}\right)\left(\frac{1-\varepsilon}{\varepsilon}\right) \tag{4.98}$$

Converting flow velocity through the mass of material particles to flow velocity through the empty reactor, then substituting into Eq. (4.98) gives

$$\begin{gathered}\Delta P = 3.5\left(\frac{\rho Z v_{\mathrm{Super}}^2}{2d_{\mathrm{S}}}\right)\left(\frac{1-\varepsilon}{\varepsilon^3}\right)\\ \Delta P = 1.75\left(\frac{\rho Z v_{\mathrm{Super}}^2}{d_{\mathrm{S}}}\right)\left(\frac{1-\varepsilon}{\varepsilon^3}\right)\end{gathered} \tag{4.99}$$

Eq. (4.99) is the Burke–Plummer equation for turbulent flow through a mass of spherical particles [11].

The Blake–Kozeny equation and the Burke–Plummer equation define the limiting flow regimes for a fluid passing through a mass of spherical particles. Ergun added these equations to obtain

$$\frac{\Delta p}{Z} = \frac{150\mu v_\infty}{d_{\mathrm{S}}^2}\frac{(1-\varepsilon)^2}{\varepsilon^3} + 1.75\left(\frac{\rho v_\infty^2}{d_{\mathrm{S}}}\right)\left(\frac{1-\varepsilon}{\varepsilon^3}\right) \tag{4.100}$$

Ergun then used published data to confirm the validity of this equation. Note that Eq. (4.100) is for spheres. For structured solids other than spheres, we modify the equation by including a "sphericity" factor Ψ, which is defined as

$$\Psi = \frac{\text{Surface area of sphere with equal volume to the nonspherical particle}}{\text{Surface area of the nonspherical particle}} \tag{4.101}$$

The Ergun equation then becomes

$$\frac{\Delta p}{Z} = \frac{150\mu v_{\infty}}{d_S^2 \Psi^2} \frac{(1-\varepsilon)^2}{\varepsilon^3} + 1.75\left(\frac{\rho v_{\infty}^2}{d_S \Psi}\right)\left(\frac{1-\varepsilon}{\varepsilon^3}\right) \tag{4.102}$$

Note that the Ergun equation

- is not based on Dimensional Analysis, Similitude, or Model Theory; therefore, it cannot be used to scale packed columns, either up or down;
- depends upon the diameter of the catalyst spheres in the packed column but not on the diameter of the column itself, except through the calculation of v_{Super};
- assumes a tortuosity of 25/12;
- assumes that $3f/2 = 3.5$;
- contains two universal constants, namely, 150 and 1.75 that actually depend upon material particle or structured solid geometry and packed column geometry [12].

In summary, it should not surprise us if the Ergun equation produces incorrect estimates of pressure difference for different shaped material particles and structured solids in the same packed column or that the Ergun equation cannot be used for scaling packed columns.

If we cannot use the Ergun equation to scale a packed column unit operation, then we must devise a different method for scaling such a unit operation. Since scaling is based on dimensionless parameters, we should base our new scaling procedure for packed columns on dimensional analysis. We can perform a dimensional analysis of fluid flow through packed columns because the variables of the process are well known and have been studied for many decades [13,14]. The variables are: fluid velocity v [LT^{-1}], column diameter D [L], characteristic length of the material mass L_{Char} that is dependent upon the size and shape of the material particles [L], pressure difference per physical height of the material column $\Delta P/Z$ [$ML^{-2}T^{-2}$], fluid density ρ [ML^{-3}], and fluid viscosity μ [$ML^{-1}T^{-1}$]. The Dimension Table for these variables is

Variables		$\Delta P/Z$	D	L_{Char}	v	ρ	μ
Dimensions	L	−2	1	1	1	−3	−1
	M	1	0	0	0	1	1
	T	−2	0	0	−1	0	−1

We define L_{Char}, the characteristic length of the material particle, as

$$L_{\text{Char}} = \frac{Z}{\varepsilon} \tag{4.103}$$

where Z, in our case, is the height of the material particle mass and ε is the void fraction of the material particles. We determine Z from the mass of material particles or structured solids charged to the column, that amount being measured by weight W_{Packing}. Dividing W_{Packing} by the loose compacted bulk density of the packing gives the volume of catalyst charged to the reactor, that is

$$\frac{W_{\text{Packing}}}{\rho_{\text{LBD}}} = V_{\text{Packing}} \tag{4.104}$$

We use loose bulk density of the material particles or structured solids rather than their compacted bulk density because most commercial packed columns and fixed-bed reactors are not shaken or vibrated during filling. Note that the volume of packing in the structural column is

$$V_{\text{Packing}} = A_{\text{CS}} Z \tag{4.105}$$

where A_{CS} is the cross-sectional area of the empty column. Dividing Eq. (4.105) by the available flow area gives

$$\frac{A_{\text{CS}} Z}{\varepsilon A_{\text{CS}}} = \frac{Z}{\varepsilon} = L_{\text{Char}} \tag{4.106}$$

Thus if $Z = 3$ m and $\varepsilon = 0.4$, then $L = 4.5$ m; or if $Z = 3$ m and $\varepsilon = 0.97$, then $L = 3.1$ m. In other words, L_{Char} is an indication of flow conduit or tube length through the packing mass. Hence, L_{Char} indicates the tortuosity of the flow path through the material or the structured solid mass.

The Dimension matrix for fluid flow through a packed column is

$$\begin{bmatrix} -2 & 1 & 1 & 1 & -3 & -1 \\ 1 & 0 & 0 & 0 & 1 & 1 \\ -2 & 0 & 0 & -1 & 0 & -1 \end{bmatrix} \tag{4.107}$$

The rank for this Dimension matrix is 3. Therefore, from Buckingham's Pi Theorem, the number of dimensionless parameters N_{P} for fluid flow through a column of material particles is

$$N_{\text{P}} = N_{\text{Var}} - R \tag{4.108}$$

where N_{Var} is the number of variables in the analysis and R is the rank of the Dimension matrix. Making the appropriate substitutions gives

$$N_P = 6 - 3 = 3 \tag{4.109}$$

Therefore, our dimensional analysis of fluid flow through a mass of material particles or through a mass of structured solids will produce three dimensionless parameters. The Total matrix for fluid flow through a catalyst mass is

$$T = \begin{array}{c} \\ \Delta P/Z \\ D \\ L_{Char} \\ v \\ \rho \\ \mu \end{array} \begin{array}{c} \begin{array}{cccccc} \Pi_1 & \Pi_2 & \Pi_3 & & & \end{array} \\ \begin{bmatrix} 1 & 0 & 0 & 0 & 0 & 0 \\ 0 & 1 & 0 & 0 & 0 & 0 \\ 0 & 0 & 1 & 0 & 0 & 0 \\ -3 & 1 & 1 & -1 & -3 & -2 \\ -2 & 1 & 1 & -1 & -2 & -1 \\ 1 & -1 & -1 & 1 & 3 & 1 \end{bmatrix} \end{array} \tag{4.110}$$

The three independent dimensionless parameters are identified above the appropriate columns of the Total matrix. The dimensionless parameters are

$$\begin{aligned} \Pi_1 &= \frac{(\Delta P/Z)\mu}{v^3\rho^2} \\ \Pi_2 &= \frac{Dv\rho}{\mu} = \mathrm{Re_D} \\ \Pi_3 &= \frac{L_{Char}v\rho}{\mu} \end{aligned} \tag{4.111}$$

where Π_1 is an unnamed dimensionless number; Π_2 is the Reynolds number based on the diameter of the solid contained in the column; Π_3 is a dimensionless number resembling the Reynolds number, but based on L_{Char}, the characteristic length of the solid contained in the column. We can multiply and/or divide these three dimensionless variables with each other because they are independent of each other. Dividing Π_3 by Π_2 gives

$$\frac{\Pi_3}{\Pi_2} = \frac{L_{Char}v\rho/\mu}{Dv\rho/\mu} = \frac{L_{Char}}{D} \tag{4.112}$$

which is an aspect ratio for the packed column. Therefore, the function describing fluid flow through a mass of material particles or structured solids is

$$f(\Pi_1, \Pi_2, \Pi_{3/2}) = 0 \tag{4.113}$$

If we identify Π_1 as the dependent parameter, then we can rewrite this function as

$$f(\Pi_2, \Pi_{3/2}) = \Pi_1 \tag{4.114}$$

Thus plotting Π_1 as a function of Π_2, the Reynolds number, produces smooth curves with parametric lines described by $\Pi_{3/2}$, which is shown in Fig. 4.1. Fig. 4.1 shows Π_1 as a function of Reynolds number for a variety of spherical particles in a number of different diameter pipes. $d(s)$ identifies sphere diameter. The curves are distinguished by the parametric $\Pi_{3/2}$, which is L_{Char}/D. Fig. 4.1 shows that Π_1 collapses to a common horizontal line in the turbulent flow regime, that is, at high Reynolds numbers. This horizontal line corresponds to the Burke–Plummer result. For low Reynolds numbers, the correlation for each size sphere is negatively sloped, which corresponds to the Blake–Kozeny result. However, unlike the Ergun equation, the different sized spheres each produce a different correlation in the laminar flow

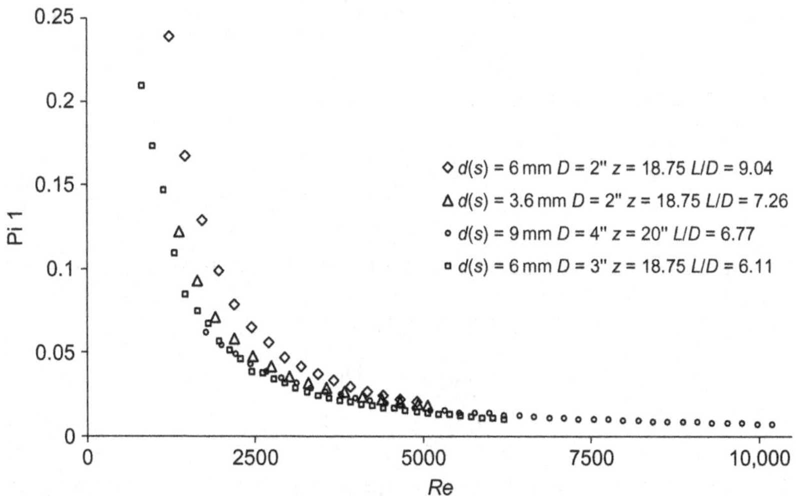

Figure 4.1 Dimensionless coefficient Π_1, or $((dP/Z)\mu)/v^3\rho^2$, as a function of Π_2, or $Dv\rho/\mu = Re_D$, with $\Pi_{3/2}$, or L/D, as parameter.

regime. Each spherical diameter also produces a different correlation in the transition flow regime. Thus the conflicting assessment with regard to the Ergun equation, namely, many engineers say it does not work while others say it does work. If fluid flows through a packed column at high Reynolds numbers, it is more likely that the Ergun equation will produce valid results. If, however, fluid flows through a packed column at low Reynolds numbers, then the Ergun equation is more likely to produce incorrect results. The same statement is valid for flow in the transition zone.

To scale a packed column, we start with Π_1 since it contains the most important variable, the pressure drop $\Delta P/Z$. The process design engineers will specify a $\Delta P/Z$ for the prototype. Note that subscript M denotes the model variables and subscript P identifies the prototype variables. The scaling criterion is

$$(\Pi_1)_\mathrm{M} = (\Pi_1)_\mathrm{P} \tag{4.115}$$

Thus

$$\frac{(\Delta P/Z)_\mathrm{M}\mu_\mathrm{M}}{v_\mathrm{M}^3\rho_\mathrm{M}^2} = \frac{(\Delta P/Z)_\mathrm{P}\mu_\mathrm{P}}{v_\mathrm{P}^3\rho_\mathrm{P}^2} \tag{4.116}$$

But, $\mu_\mathrm{M} = \mu_\mathrm{P}$ and $\rho_\mathrm{M} = \rho_\mathrm{P}$; therefore, Eq. (4.116) becomes

$$\frac{(\Delta P/Z)_\mathrm{M}}{v_\mathrm{M}^3} = \frac{(\Delta P/Z)_\mathrm{P}}{v_\mathrm{P}^3} \tag{4.117}$$

Solving the above relationship for v_P^3 yields

$$v_\mathrm{P} = \left[\frac{(\Delta P/Z)_\mathrm{M}}{(\Delta P/Z)_\mathrm{P}}\right]^{1/3} v_\mathrm{M} \tag{4.118}$$

We next determine the diameter of the packed column using Π_2. Thus

$$(\Pi_1)_\mathrm{M} = (\Pi_1)_\mathrm{P}$$

$$\frac{D_\mathrm{M}v_\mathrm{M}\rho_\mathrm{M}}{\mu_\mathrm{M}} = \frac{D_\mathrm{P}v_\mathrm{P}\rho_\mathrm{P}}{\mu_\mathrm{P}} \tag{4.119}$$

But, as above $\mu_\mathrm{M} = \mu_\mathrm{P}$ and $\rho_\mathrm{M} = \rho_\mathrm{P}$; therefore, Eq. (4.119) becomes

$$D_\mathrm{M}v_\mathrm{M} = D_\mathrm{P}v_\mathrm{P} \tag{4.120}$$

We know D_M and v_M from our model and we just calculated v_P above; thus

$$D_P = \left(\frac{v_M}{v_P}\right) D_M \tag{4.121}$$

To determine the height of the column and the weight of material particles or structured solids with which to charge it, we use the dimensionless parameter $\Pi_{3/2}$. As before

$$\begin{aligned}(\Pi_{3/2})_M &= (\Pi_{3/2})_P \\ \frac{L_{Char,M}}{D_M} &= \frac{L_{Char,P}}{D_P}\end{aligned} \tag{4.122}$$

We know D_M and $L_{Char,M}$ from our model and we calculated D_P above, so

$$L_{Char,P} = \left(\frac{D_P}{D_M}\right) L_{Char,M} \tag{4.123}$$

We convert $L_{Char,P}$ to Z_P, the physical height of the material particles or structured solids comprising the packing mass by

$$Z_P = \varepsilon L_{Char,P} \tag{4.124}$$

Remember, Z_P is the height of the physical packing in the column. Therefore, the height of the containing column must be greater than Z_P.

We purchase material particles or structured solids on a weight basis. Therefore, we need to convert Z_P to weight. The volume of packing yielding height Z_P is

$$V_P = \left(\frac{\pi}{4}\right) D_P^2 L_{Char,P} \tag{4.125}$$

where V_P is the material volume of the mass in the prototype column. Multiplying V_P by loose bulk density ρ_{LBD} gives the weight of packing contained in the prototype column; thus

$$W_P = \rho_{LBD} V_P = \left(\frac{\rho_{LBD}\pi}{4}\right) D_P^2 L_{Char,P} \tag{4.126}$$

The method based on dimensional analysis produces curves similar to those presented by Ergun in 1952 and by myriad others since 1952.

Unfortunately, we cannot use such information for scaling a packed column unit operation since they do not contain process variables such as column diameter and flow path length. However, the method based on dimensional analysis does contain process related variables. And, since model theory is based on dimensional analysis, we can use model theory to scale a packed column unit operation, as we did earlier.

SUMMARY

This chapter applied dimensional analysis to scaling operations and processes requiring mechanical similarity. Since mechanical similarity only requires geometric similarity and depends upon three dimensions only, our scaling efforts are, for the most part, exact. In other words, we do not require a distorted model when scaling to a prototype. Also, we have only three control regimes to consider in mechanical similarity: laminar flow, transition flow, and turbulent flow. When scaling fluid processes, we must identify which control regime controls the model. Whichever control regime controls the model will be the one we scale to the prototype, unless we distort the model. *Note*: Never scale from the transition flow regime or into the transition flow regime—the outcome in either case will most likely not be the desired result.

REFERENCES

[1] L. Moody, Friction factors for pipe flow, Trans. Am. Soc. Mech. Eng. 66 (1944) 67.

[2] J. Hunsaker, B.N. Rightmere, Engineering Applications of Fluid Mechanics, McGraw-Hill Book Company, New York, NY, 1947, pp. 126–127.

[3] G. Murphy, Similitude in Engineering, The Ronald Press Company, New York, NY, 1950. p. 140.

[4] R. Johnstone, M. Thring, Pilot Plants, Models, and Scale-up Methods in Chemical Engineering, McGraw-Hill Book Company, Inc, New York, NY, 1957. p. 115.

[5] S. Ergun, Fluid flow through packed columns, Chem. Eng. Progress 48 (1952) 89.

[6] L. Sissom, D. Pitts, Elements of Transport Phenomena, McGraw-Hill Book Company, Inc, New York, NY, 1972 (Chapter 20).

[7] W. Janna, Introduction to Fluid Mechanics, Brooks/Cole Engineering Division (Wadsworth, Inc.), Monterey, CA, 1983. p. 151.

[8] W. McCabe, J. Smith, Unit Operations of Chemical Engineering, third ed., McGraw-Hill, Inc, 1976. p. 148.

[9] N. de Nevers, Fluid Mechanics for Chemical Engineers, second ed., McGraw-Hill, New York, NY, 1991. p. 413.

[10] F. Blake, The resistance of packing to fluid flow, Trans. Am. Inst. Chem. Eng. 14 (1922) 415.

[11] S. Burke, W. Plummer, Gas flow through packed beds, Ind. Eng. Chem. 20 (1928) 1196.

[12] D. Nemec, J. Levec, Flow through packed bed reactors: 2. Two phase concurrent downflow, Chem. Eng. Sci. 60 (2005) 694.

[13] J. Worstell, Dimensional Analysis: Practical Guides in Chemical Engineering, Butterworth-Heinemann, Oxford, UK, 2014.

[14] J. Worstell, Adiabatic Fixed-Bed Reactors: Practical Guides in Chemical Engineering, Butterworth-Heinemann, Oxford, UK, 2014.

CHAPTER 5

Scaling Heat Transfer

INTRODUCTION

Nearly all chemical processes involve the consumption or generation of energy. This energy moves about the chemical process as heat. The mathematical description of heat flow requires a fourth dimension, which is temperature. We use temperature because heat moves from regions of higher temperature to regions of lower temperature. This additional dimension increases the complexity of scaling a chemical process, either to a larger size or to a smaller size.

HEATED, UNINSULATED STORAGE TANK

Let us assume that we work for a drilling mud formulator with a manufacturing site located on the Gulf coast of the United States. Our CEO wants to expand operations to the Bakken shale gas reservoir in North Dakota. Some of the chemicals that go into the drilling mud have high melting points; therefore, we store them in horizontal, cylindrical, heated tanks. These storage tanks are not insulated due to "corrosion under insulation" issues that occur along the humid Gulf coast. We do build cement block walls around these tanks to reduce energy loss by convective heat transfer. Each such tank contains an internal coil through which heating medium flows. A temperature controller throttles the flow of heating medium through this internal coil. The temperature of the heating medium is T_{HM} and the temperature of the stored fluid is T_{SF}. ΔT_{HMSF} is $T_{HM} - T_{SF}$ and its dimension is [θ]. The energy lost, Q_{Lost} $[L^2MT^{-3}]$ from the tank is

$$Q_{Lost} = UA(\Delta T_{TA}) \tag{5.1}$$

where U is the overall heat transfer coefficient $[MT^{-3}\theta^{-1}]$; A is the area through which the heat moves $[L^2]$; and ΔT_{TA} is the difference between the temperature of the stored fluid and atmospheric temperature [θ]. The temperature controller ensures that $Q_{In} = Q_{Lost}$,

© 2016 Elsevier Inc. All rights reserved.

Table 5.1 Dimension Table for Scaling an Uninsulated, Internally Heated Storage Vessel

	D	L	L_C	$C_{P,SF}$	k_{SF}	μ_{SF}	ρ_{SF}	UA	ΔT_{HMSF}	ΔT_{VA}	$C_{P,HM}$	k_{HM}	μ_{HM}	ρ_{HM}	D_C	v	h
L	1	1	1	2	1	−1	−3	0	0	0	2	1	−1	−3	1	1	0
M	0	0	0	0	1	1	1	1	0	0	0	1	1	1	0	0	1
T	0	0	0	−2	−3	−1	0	−3	0	0	−2	−3	−1	0	0	−1	−3
θ	0	0	0	−1	−1	0	0	−1	1	1	−1	−1	0	0	0	0	−1

where Q_{In} is the heat supplied by the internal coil to the stored fluid. Our CEO foresees a larger size facility in North Dakota than the one on the Gulf coast. Therefore, we need to scale these uninsulated, but sheltered storage tanks.

The geometric variables are tank diameter D [L], tank length [L], internal coil length L_C [L], and internal coil diameter D_C [L]. The material variables are the heat capacity of the heating medium $C_{P,HM}$ and the heat capacity of the stored fluid $C_{P,SF}$, both $[L^2T^{-2}\theta^{-1}]$, the heat conductivity of the heating medium k_{HM} and the heat conductivity of the stored fluid k_{SF}, both $[LMT^{-3}\theta^{-1}]$, the viscosity of the heating medium μ_{HM} and the viscosity of the stored fluid μ_{SF}, both $[L^{-1}MT^{-1}]$, and the density of the heating medium ρ_{HM} and the density of the stored fluid ρ_{SF}, both $[ML^{-3}]$. The process variables are heating medium fluid velocity v_{HF} $[LT^{-1}]$, the temperature difference between the heating medium and the stored fluid ΔT_{HMSF} [θ], the temperature difference between the tank and the atmosphere ΔT_{TA} [θ], the overall heat transfer coefficient for heat lost to the atmosphere from the tank UA $[MT^{-3}\theta^{-1}]$, and the heat transfer coefficient for the internal coil h_{HF} $[MT^{-3}\theta^{-1}]$. We assume that heat conductivity through each metal surface is negligible.

Table 5.1 contains the Dimension Table for this scaling exercise. From its Dimension matrix, we identify its Rank matrix as

$$R=\begin{bmatrix} -3 & 1 & 1 & 0 \\ 1 & 0 & 0 & 1 \\ 0 & 0 & -1 & -3 \\ 0 & 0 & 0 & -1 \end{bmatrix} \tag{5.2}$$

The determinant of R is -1; therefore, it is nonsingular and its Rank is 4. The number of dimensionless parameters is

$$N_P = N_{Var} - R = 17 - 4 = 13 \quad (5.3)$$

The inverse of the Rank matrix is

$$R^{-1} = \begin{bmatrix} 0 & 1 & 0 & 1 \\ 1 & 3 & 1 & 0 \\ 0 & 0 & -1 & 3 \\ 0 & 0 & 0 & -1 \end{bmatrix} \quad (5.4)$$

The Bulk matrix is

$$B = \begin{bmatrix} 1 & 1 & 1 & 2 & 1 & -1 & -3 & 0 & 0 & 0 & 2 & 1 & -1 \\ 0 & 0 & 0 & 0 & 1 & 1 & 1 & 1 & 0 & 0 & 0 & 1 & 1 \\ 0 & 0 & 0 & -2 & -3 & -1 & 0 & -3 & 0 & 0 & -2 & -3 & -1 \\ 0 & 0 & 0 & -1 & -1 & 0 & 0 & -1 & 1 & 1 & -1 & -1 & 0 \end{bmatrix} \quad (5.5)$$

The result of multiplying the $-R^{-1}$ by the B matrix is

$$-R^{-1} \cdot B = \begin{bmatrix} 0 & 0 & 0 & 1 & 0 & -1 & -1 & 0 & -1 & -1 & 1 & 0 & -1 \\ -1 & -1 & -1 & 0 & -1 & -1 & 0 & 0 & 0 & 0 & 0 & -1 & -1 \\ 0 & 0 & 0 & 1 & 0 & -1 & 0 & 0 & -3 & -3 & 1 & 0 & -1 \\ 0 & 0 & 0 & -1 & -1 & 0 & 0 & -1 & 1 & 1 & -1 & -1 & 0 \end{bmatrix} \quad (5.6)$$

We can now assemble the Total matrix for scaling a heated, uninsulated storage tank; it is shown in Total Matrix 1. We list the variables of this scaling effort along the left side of the Total matrix. We identify the dimensionless parameters above each of the appropriate columns of the Total matrix. The resulting dimensionless parameters are

$$\Pi_1 = \frac{D}{D_C} \quad \Pi_2 = \frac{L}{D_C} \quad \Pi_3 = \frac{L_C}{D_C} \quad \Pi_4 = \frac{C_{P,SF}\rho_{HM}v_{HM}}{h_{HM}}$$

$$\Pi_5 = \frac{k_{SF}}{D_C h_{HM}} \quad \Pi_6 = \frac{\mu_{SF}}{\rho_{HM} D_C v_{HM}} \quad \Pi_7 = \frac{\rho_{SF}}{\rho_{HM}} \quad \Pi_8 = \frac{UA}{h_{HM}}$$

$$\Pi_9 = \frac{h_{HM}}{\rho_{HM} v_{HM}^3 \Delta T_{SFHM}} \quad \Pi_{10} = \frac{h_{HM}}{\rho_{HM} v_{HM}^3 \Delta T_{TA}} \quad \Pi_{11} = \frac{C_{P,HM}\rho_{HM}v_{HM}}{h_{HM}} \quad \Pi_{12} = \frac{k_{HM}}{D_C h_{HM}}$$

$$\Pi_{13} = \frac{\mu_{HM}}{\rho_{HM} D_C v_{HM}}$$

Total Matrix 1 The Total Matrix for Scaling an Uninsulated, Internally Heated Storage Vessel

	Π_1	Π_2	Π_3	Π_4	Π_5	Π_6	Π_7	Π_8	Π_9	Π_{10}	Π_{11}	Π_{12}	Π_{13}				
D	1	0	0	0	0	0	0	0	0	0	0	0	0	0	0	0	0
L	0	1	0	0	0	0	0	0	0	0	0	0	0	0	0	0	0
L_C	0	0	1	0	0	0	0	0	0	0	0	0	0	0	0	0	0
$C_{P,SF}$	0	0	0	1	0	0	0	0	0	0	0	0	0	0	0	0	0
k_{SF}	0	0	0	0	1	0	0	0	0	0	0	0	0	0	0	0	0
μ_{SF}	0	0	0	0	0	1	0	0	0	0	0	0	0	0	0	0	0
ρ_{SF}	0	0	0	0	0	0	1	0	0	0	0	0	0	0	0	0	0
UA	0	0	0	0	0	0	0	1	0	0	0	0	0	0	0	0	0
$T = \Delta T_{HMSF}$	0	0	0	0	0	0	0	0	1	0	0	0	0	0	0	0	0
ΔT_{VA}	0	0	0	0	0	0	0	0	0	1	0	0	0	0	0	0	0
$C_{P,HF}$	0	0	0	0	0	0	0	0	0	0	1	0	0	0	0	0	0
k_{HF}	0	0	0	0	0	0	0	0	0	0	0	1	0	0	0	0	0
μ_{HF}	0	0	0	0	0	0	0	0	0	0	0	0	1	0	0	0	0
ρ_{HF}	0	0	0	1	0	−1	−1	0	−1	−1	1	0	−1	0	1	0	1
D_C	−1	−1	−1	0	−1	−1	0	0	0	0	0	−1	−1	1	3	1	0
ν_{hf}	0	0	0	1	0	−1	0	0	−3	−3	1	0	−1	0	0	−1	3
h_{hf}	0	0	0	−1	−1	0	0	−1	1	1	−1	−1	0	0	0	0	−1

Some of the above dimensionless parameters are recognizable; for example, Π_1, Π_2, and Π_3 are geometric ratios for the tank and coil, Π_{12} is the inverse Nusselt number in terms of coil diameter for the heating medium, and Π_{13} is the inverse Reynolds number for the heating medium.

Note that many of the above dimensionless parameters contain variables from the stored fluid and from the heating medium. Such dimensionless parameters are meaningless. Thus our first task is to assemble a set of dimensionless parameters in which each parameter is comprised of either stored fluid variables or heating medium variables. For example, dividing Π_5 by the product of $\Pi_1\Pi_8$ provides us with an inverse Nusselt number for the stored fluid

$$\frac{\Pi_5}{\Pi_1\Pi_8} = \frac{k_{SF}}{D(UA)} = Nu_{SF}^{-1} \tag{5.7}$$

and

$$\frac{\Pi_4\Pi_6}{\Pi_5} = \frac{C_{P,SF}\mu_{SF}}{k_{SF}} = Pr_{SF} \quad \text{and} \quad \frac{\Pi_{11}\Pi_{13}}{\Pi_{12}} = \frac{C_{P,HM}\mu_{HM}}{k_{HM}} = Pr_{HM} \tag{5.8}$$

are the Prandtl numbers for the stored fluid and the heating medium, respectively. Dividing Π_{11} by Π_{12} gives us the Peclet number for the heating medium

$$\frac{\Pi_{11}}{\Pi_{12}} = \frac{C_{P,HM}\rho_{HM}v_{HM}D_C}{k_{HM}} = Pe_{HM} \tag{5.9}$$

Other derived dimensionless parameters are

$$\frac{\Pi_2}{\Pi_3} = \frac{L}{L_C} \qquad \frac{\Pi_9}{\Pi_{10}} = \frac{\Delta T_{TA}}{\Delta T_{HMSF}} \qquad \frac{\Pi_6}{\Pi_{13}} = \frac{\mu_{SF}}{\mu_{HN}}$$

Π_{12} is the inverse Nusselt number for heat transfer to or from the coil. It is based on coil diameter. However, heat transfer to or from the coil occurs along its length. Therefore, we should express the Nusselt number in terms of coil length. We do that by dividing Π_{12} by Π_3, which gives us

$$\frac{\Pi_{12}}{\Pi_3} = \frac{k_{HM}/D_C h_{HM}}{L_C/D_C} = \frac{k_{HM}}{L_C h_{HM}} \tag{5.10}$$

The thirteen dimensionless parameters required to scale an internally heated, uninsulated storage tank are

$$\Pi_1 = \frac{D}{D_C} \qquad \frac{\Pi_2}{\Pi_3} = \frac{L}{L_C} \qquad \Pi_3 = \frac{L_C}{D_C} \qquad \Pi_7 = \frac{\rho_{SF}}{\rho_{HM}}$$

$$\Pi_8 = \frac{UA}{h_{HM}} \qquad \frac{\Pi_6}{\Pi_{13}} = \frac{\mu_{SF}}{\mu_{HN}} \qquad \frac{\Pi_{12}}{\Pi_3} = \frac{k_{HM}}{L_C h_{HM}} = Nu_{HM}^{-1} \qquad \Pi_{13} = \frac{\mu_{HM}}{\rho_{HM} D_C v_{HM}} = Re_{HM}^{-1}$$

$$\frac{\Pi_5}{\Pi_1\Pi_8} = \frac{k_{SF}}{D(UA)} = Nu_{SF}^{-1} \qquad \frac{\Pi_4\Pi_6}{\Pi_5} = \frac{C_{P,SF}\mu_{SF}}{k_{SF}} = Pr_{SF} \qquad \frac{\Pi_{11}\Pi_{13}}{\Pi_{12}} = \frac{C_{P,HM}\mu_{HM}}{k_{HM}} = Pr_{HM} \qquad \frac{\Pi_{11}}{\Pi_{12}} = \frac{C_{P,HM}\rho_{HM}v_{HM}D_C}{k_{HM}} = Pe_{HM}$$

$$\frac{\Pi_9}{\Pi_{10}} = \frac{\Delta T_{TA}}{\Delta T_{HMSF}}$$

Those dimensionless parameters that are ratios of material variables, such as Π_7, which is the ratio of fluid densities, and Π_6/Π_{13}, which is the ratio of fluid viscosities, are not an issue for this scaling effort. The criteria for scaling these dimensionless parameters are $(\Pi_7)_M = (\Pi_7)_P$ and $(\Pi_6/\Pi_{13})_M = (\Pi_6/\Pi_{13})_P$, where the subscript M denotes model and the subscript P denotes prototype. These criteria must hold true. The other combinations of material variables are also subject to the above restriction; in other words, they do not complicate the scaling effort unless we change either the stored fluid or the heating medium, thus

$$\left(\frac{\Pi_4\Pi_6}{\Pi_5}\right)_M = \left(\frac{\Pi_4\Pi_6}{\Pi_5}\right)_E \quad \text{and} \quad \left(\frac{\Pi_{11}\Pi_{13}}{\Pi_{12}}\right)_M = \left(\frac{\Pi_{11}\Pi_{13}}{\Pi_{12}}\right)_P \tag{5.11}$$

must hold true. The ratios $\Pi_4\Pi_6/\Pi_5$ and $\Pi_{11}\Pi_{13}/\Pi_{12}$ are the Prandtl numbers for the stored fluid and the heating medium, respectively.

If for some reason we are not able to meet the requirement that all the model dimensionless parameters equal their appropriate prototype dimensionless parameter, then we must determine which dimensionless parameters are the most important for our scaling effort and use them and neglect the nonconforming dimensionless parameters. All such choices and the reasoning behind each choice must be documented for future use.

Note that the above set of dimensionless parameters requires the independent calculation of U and h_{HM}. As stated earlier, U is the overall heat transfer coefficient for heat lost from the storage vessel to the atmosphere. It is the sum of convective and radiative heat transfer mechanisms. We quantify these heat transfer mechanisms with empirical relationships. For the convective mechanism, we use

$$h_{Conv} = \kappa(\Delta T_{TA})^{0.25} \tag{5.12}$$

where h_{Conv} is the convection heat transfer coefficient $[MT^{-3}\theta^{-1}]$ and κ is a constant between 0.2 and 0.3—it has dimensions appropriate for h_{Conv}. For the radiative mechanism, we use

$$h_{Rad} = \varepsilon\sigma\left[\frac{(T_T)^4 - (T_{Atm})^4}{T_T - T_{Atm}}\right] \tag{5.13}$$

where h_{Rad} is the radiation heat transfer coefficient $[MT^{-3}\theta^{-1}]$; ε is the emissivity of the metal surface; σ is the Stefan–Boltzmann constant; and T_T and T_{Atm} are the tank temperature and the atmospheric temperature, respectively [1].

To determine the value for h_{HM}, we again use dimensional analysis [2]. Consider a fluid flowing inside a pipe or tube. The heat transferred to the flowing fluid is

$$q = hA\,\Delta T\,\Delta t \tag{5.14}$$

where q is the heat transferred to the flowing fluid [L^2MT^{-2}]; h is the convective heat transfer coefficient for the process [$MT^{-3}\theta$]; A is the area through which heat is transferred [L^2]; ΔT is the temperature difference between the pipe's wall temperature and the average temperature of the flowing fluid, given by $T_{Ave} = (T_{CenterLine} - T_{Wall})/2$, both [$\theta$]; and Δt is the time duration of heat flow [T].

The geometric variable is pipe diameter D [L]. The material variables are fluid density ρ [$L^{-3}M$], fluid dynamic viscosity μ [$L^{-1}MT^{-1}$], fluid heat capacity C_P [$L^2MT^{-2}\theta^{-1}$], fluid thermal conductivity k [$LMT^{-3}\theta^{-1}$], and fluid heat transfer coefficient h [$MT^{-3}\theta^{-1}$]. The process variables are fluid velocity v [LT^{-1}], average fluid temperature T_{Ave} [θ], and pipe wall temperature T_{Pipe} [θ]. The Dimensional Table is

Variable		T_{Ave}	T_{Pipe}	D	v	μ	ρ	C_P	k	h
Dimension	L	0	0	1	1	−1	−3	2	1	0
	M	0	0	0	0	1	1	0	1	1
	T	0	0	0	−1	−1	0	−2	−3	−3
	θ	1	1	0	0	0	0	−1	−1	−1

from which we write the Dimension matrix, which is

$$\begin{bmatrix} 0 & 0 & 1 & 1 & -1 & -3 & 2 & 1 & 0 \\ 0 & 0 & 0 & 0 & 1 & 1 & 0 & 1 & 1 \\ 0 & 0 & 0 & -1 & -1 & 0 & -2 & -3 & -3 \\ 1 & 1 & 0 & 0 & 0 & 0 & -1 & -1 & -1 \end{bmatrix} \tag{5.15}$$

The largest square matrix for this Dimension matrix is 4×4; it is

$$R = \begin{bmatrix} -3 & 2 & 1 & 0 \\ 1 & 0 & 1 & 1 \\ 0 & -2 & -3 & -3 \\ 0 & -1 & -1 & -1 \end{bmatrix} \tag{5.16}$$

Its determinant is

$$|R| = \begin{vmatrix} -3 & 2 & 1 & 0 \\ 1 & 0 & 1 & 1 \\ 0 & -2 & -3 & -3 \\ 0 & -1 & -1 & -1 \end{vmatrix} = -1 \tag{5.17}$$

The determinant for the Rank matrix is nonsingular; thus, the Rank of this Dimension matrix is 4. The number of dimensionless parameters is

$$N_P = N_{Var} - R = 9 - 4 = 5 \tag{5.18}$$

The inverse of R is

$$R^{-1} = \begin{bmatrix} -3 & 2 & 1 & 0 \\ 1 & 0 & 1 & 1 \\ 0 & -2 & -3 & -3 \\ 0 & -1 & -1 & -1 \end{bmatrix}^{-1} = \begin{bmatrix} 0 & 1 & 1 & -2 \\ 0 & 0 & 1 & -3 \\ 1 & 3 & 1 & 0 \\ -1 & -3 & -2 & 2 \end{bmatrix} \tag{5.19}$$

The Bulk matrix is

$$B = \begin{bmatrix} 0 & 0 & 1 & 1 & -1 \\ 0 & 0 & 0 & 0 & 1 \\ 0 & 0 & 0 & -1 & -1 \\ 1 & 1 & 0 & 0 & 0 \end{bmatrix} \tag{5.20}$$

and $-R^{-1} \cdot B$ is

$$-R^{-1} \cdot B = -\left[\begin{bmatrix} 0 & 1 & 1 & -2 \\ 0 & 0 & 1 & -3 \\ 1 & 3 & 1 & 0 \\ -1 & -3 & -2 & 2 \end{bmatrix}\begin{bmatrix} 0 & 0 & 1 & 1 & -1 \\ 0 & 0 & 0 & 0 & 1 \\ 0 & 0 & 0 & -1 & -1 \\ 1 & 1 & 0 & 0 & 0 \end{bmatrix}\right] = \begin{bmatrix} 2 & 2 & 0 & 1 & 0 \\ 3 & 3 & 0 & 1 & 1 \\ 0 & 0 & -1 & 0 & -1 \\ -2 & -2 & 1 & -1 & 0 \end{bmatrix} \tag{5.21}$$

We can now complete the Total matrix, which is

$$T = \begin{array}{c} \\ K_{Ave} \\ K_{Pipe} \\ D \\ v \\ \mu \\ \rho \\ C_P \\ k \\ h \end{array} \begin{array}{c} \begin{array}{ccccccccc} \Pi_1 & \Pi_2 & \Pi_3 & \Pi_4 & \Pi_5 & & & & \end{array} \\ \begin{bmatrix} 1 & 0 & 0 & 0 & 0 & 0 & 0 & 0 & 0 \\ 0 & 1 & 0 & 0 & 0 & 0 & 0 & 0 & 0 \\ 0 & 0 & 1 & 0 & 0 & 0 & 0 & 0 & 0 \\ 0 & 0 & 0 & 1 & 0 & 0 & 0 & 0 & 0 \\ 0 & 0 & 0 & 0 & 1 & 0 & 0 & 0 & 0 \\ 2 & 2 & 0 & 1 & 0 & 0 & 1 & 1 & -2 \\ 3 & 3 & 0 & 1 & 1 & 0 & 0 & 1 & -3 \\ 0 & 0 & -1 & 0 & -1 & 1 & 3 & 1 & 0 \\ -2 & -2 & 1 & -1 & 0 & -1 & -3 & -2 & 2 \end{bmatrix} \end{array} \tag{5.22}$$

The dimensionless parameters are, reading down each Π_i column of the Total matrix

$$\Pi_1 = \frac{T_{\text{Ave}}\rho^2 C_P^3}{h^2} \quad \Pi_2 = \frac{T_{\text{Pipe}}\rho^2 C_P^3}{h^2} \quad \Pi_3 = \frac{hD}{k} = Nu \quad \Pi_4 = \frac{\rho v C_P}{h} \quad \Pi_5 = \frac{\mu C_P}{k} = Pr$$

Manipulating the above dimensionless parameters in order to obtain more recognizable dimensionless numbers, we get

$$\frac{\Pi_1}{\Pi_2} = \frac{T_{\text{Ave}}\rho^2 C_P^3/h^2}{T_{\text{Pipe}}\rho^2 C_P^3/h^2} = \frac{T_{\text{Ave}}}{T_{\text{Pipe}}} \qquad \frac{\Pi_3\Pi_4}{\Pi_5} = \frac{(hD/k)(\rho v C_P/h)}{\mu C_P/k} = \frac{\rho D v}{\mu} = Re \qquad \frac{\Pi_4^2}{\Pi_1} = \frac{(\rho v C_P/h)^2}{T_{\text{Ave}}\rho^2 C_P^3/h^2} = \frac{v^2}{T_{\text{Ave}} C_P}$$

Thus the solution is

$$f\left(\Pi_3, \Pi_5, \frac{\Pi_3\Pi_4}{\Pi_5}, \frac{\Pi_1}{\Pi_2}, \frac{\Pi_4^2}{\Pi_1}\right) = f\left(Nu, Pr, Re, \frac{\Pi_1}{\Pi_2}, \frac{\Pi_4^2}{\Pi_1}\right) = 0 \tag{5.23}$$

In terms of the heat transfer coefficient, the solution is

$$Nu = \kappa \cdot f\left(Pr, Re, \frac{\Pi_1}{\Pi_2}, \frac{\Pi_4^2}{\Pi_1}\right) \tag{5.24}$$

where κ is a constant. The fourth and fifth terms in these functions are generally considered negligible when analyzing such processes. Thus

$$Nu = \kappa \cdot f(Pr, \text{Re},) \tag{5.25}$$

We generally assume the function to be a power law relationship of the form

$$Nu = \kappa \cdot (Re)^x (Pr)^y \tag{5.26}$$

where x and y are constants determined experimentally. For turbulent flow, x lies between 0.5 and 0.8 and y lies between 0.3 and 0.4; for laminar flow x is 0.3 and y is 0.3 [3–5].

HEATED, INSULATED BATCH MIXER

Consider an insulated vessel equipped with an overhead agitator. The vessel contains an internal coil through which a heating medium flows.

We periodically fill this vessel with a fluid at temperature T_{In}, heat the fluid to temperature T_{Out}, then discharge the heated fluid to the "downstream" chemical process. The time to heat is Δt. Our next production facility will be considerably larger than our current facility. We plan to install a geometrically similar vessel in our new facility. Therefore, we need to design a thermally similar vessel for that facility.

The geometric variables are vessel diameter D [L], vessel length L [L], heating coil diameter D_{C} [L], heating coil length L_{C} [L], and agitator blade length L_{B} [L]. The material variables for the heated fluid inside the vessel are heat capacity C_{P} [$\text{L}^2\text{T}^{-2}\theta^{-1}$], heat conductivity [$\text{LMT}^{-3}\theta^{-1}$], convective heat transfer coefficient h [$\text{MT}^{-3}\theta^{-1}$], bulk fluid viscosity μ [$\text{L}^{-1}\text{MT}^{-1}$], wall fluid viscosity μ_{W} [$\text{L}^{-1}\text{MT}^{-1}$], and fluid density ρ [ML^{-3}]. The material variables for the heating medium are heat capacity $C_{\text{P,HM}}$ [$\text{L}^2\text{T}^{-2}\theta^{-1}$], heat conductivity k_{HM} [$\text{LMT}^{-3}\theta^{-1}$], convective heat transfer coefficient h_{HM} [$\text{MT}^{-3}\theta^{-1}$], viscosity μ_{HM} [$\text{L}^{-1}\text{MT}^{-1}$], and fluid density ρ_{HM} [ML^{-3}]. The process variables are agitator rotating speed N [T^{-1}], time to reach target temperature Δt [T], which is $t_{\text{Out}} - t_{\text{In}}$, heating medium fluid velocity v_{HF} [LT^{-1}], and the temperature difference for the process ΔT [θ], which is $T_{\text{Out}} - T_{\text{In}}$.

Table 5.2 contains the Dimension Table for this scaling exercise. From the Dimension matrix, which we obtain from the Dimension Table, we get the Rank matrix, which is

$$R = \begin{bmatrix} -3 & 1 & 1 & 0 \\ 1 & 0 & 0 & 1 \\ 0 & 0 & -1 & -3 \\ 0 & 0 & 0 & -1 \end{bmatrix} \tag{5.27}$$

The determinant of R is -1; therefore, it is nonsingular and its Rank is 4. The number of dimensionless parameters is

$$N_{\text{P}} = N_{\text{Var}} - R = 20 - 4 = 16 \tag{5.28}$$

Table 5.2 Dimension Table for Scaling an Insulated, Internally Heated Mixing Vessel

	D	L	L_{C}	L_{B}	C_{P}	k	h	μ	μ_{W}	ρ	ΔT	Δt	N	$C_{\text{P,HM}}$	k_{HM}	μ_{HM}	ρ_{HM}	D_{C}	v_{HM}	h_{HM}
L	1	1	1	1	2	1	0	−1	−1	−3	0	0	0	2	1	−1	−3	1	1	0
M	0	0	0	0	0	1	1	1	1	1	0	0	0	0	1	1	1	0	0	1
T	0	0	0	0	−2	−3	−3	−1	−1	0	0	1	−1	−2	−3	−1	0	0	−1	−3
θ	0	0	0	0	−1	−1	−1	0	0	0	1	0	0	−1	−1	0	0	0	0	−1

The inverse of the Rank matrix is

$$R^{-1}=\begin{bmatrix}0 & 1 & 0 & 1\\ 1 & 3 & 1 & 0\\ 0 & 0 & -1 & 3\\ 0 & 0 & 0 & -1\end{bmatrix} \tag{5.29}$$

The Bulk matrix is

$$B=\begin{bmatrix}1 & 1 & 1 & 1 & 2 & 1 & 0 & -1 & -1 & -3 & 0 & 0 & 0 & 2 & 1 & -1\\ 0 & 0 & 0 & 0 & 0 & 1 & 1 & 1 & 1 & 1 & 0 & 0 & 0 & 0 & 1 & 1\\ 0 & 0 & 0 & 0 & -2 & -3 & -3 & -1 & -1 & 0 & 0 & 1 & -1 & -2 & -3 & -1\\ 0 & 0 & 0 & 0 & -1 & -1 & -1 & 0 & 0 & 0 & 1 & 0 & 0 & -1 & -1 & 0\end{bmatrix} \tag{5.30}$$

Thus, the product of $-R^{-1}$ multiplied by the B matrix is

$$-R^{-1}\cdot B=\begin{bmatrix}0 & 0 & 0 & 0 & 1 & 0 & 0 & -1 & -1 & -1 & -1 & 0 & 0 & 1 & 0 & -1\\ -1 & -1 & -1 & -1 & 0 & -1 & 0 & -1 & -1 & 0 & 0 & -1 & 1 & 0 & -1 & -1\\ 0 & 0 & 0 & 0 & 1 & 0 & 0 & -1 & -1 & 0 & -3 & 1 & -1 & 1 & 0 & -1\\ 0 & 0 & 0 & 0 & -1 & -1 & -1 & 0 & 0 & 0 & 1 & 0 & 0 & -1 & -1 & 0\end{bmatrix} \tag{5.31}$$

We can now assemble the Total matrix for scaling a heated, insulated batch mixer; it is shown in Total Matrix 2. We list the variables of this scaling effort along the left border of the Total matrix, at the beginning of the appropriate row. We identify the dimensionless parameters above each of the appropriate columns of the Total matrix. The dimensionless parameters are

$$\Pi_1=\frac{D}{D_C} \qquad \Pi_2=\frac{L}{D_C} \qquad \Pi_3=\frac{L_C}{D_C} \qquad \Pi_4=\frac{L_B}{D_C}$$

$$\Pi_5=\frac{C_P\rho_{HM}v_{HM}}{h_{HM}} \qquad \Pi_6=\frac{k}{D_C h_{HM}} \qquad \Pi_7=\frac{h}{h_{HM}} \qquad \Pi_8=\frac{\mu}{\rho_{HM}D_C v_{HM}}$$

$$\Pi_9=\frac{\mu_W}{\rho_{HM}D_C v_{HM}} \qquad \Pi_{10}=\frac{\rho}{\rho_{HM}} \qquad \Pi_{11}=\frac{h_{HM}\Delta T}{\rho_{HM}v_{HM}^3} \qquad \Pi_{12}=\frac{v_{HM}\Delta t}{D_C}$$

$$\Pi_{13}=\frac{D_C N}{v_{HM}} \qquad \Pi_{14}=\frac{C_{P,HM}\rho_{HM}v_{HM}}{h_{HM}} \qquad \Pi_{15}=\frac{k_{HM}}{D_C h_{HM}}=Nu_{HM}^{-1} \qquad \Pi_{16}=\frac{\mu_{HM}}{\rho_{HM}D_C v_{HM}}=Re_{HM}^{-1}$$

Total Matrix 2 The Total Matrix for Scaling an Insulated, Internally Heated Storage Vessel

$T =$

	Π_1	Π_2	Π_3	Π_4	Π_5	Π_6	Π_7	Π_8	Π_9	Π_{10}	Π_{11}	Π_{12}	Π_{13}	Π_{14}	Π_{15}	Π_{16}				
D	1	0	0	0	0	0	0	0	0	0	0	0	0	0	0	0	0	0	0	0
L	0	1	0	0	0	0	0	0	0	0	0	0	0	0	0	0	0	0	0	0
L_C	0	0	1	0	0	0	0	0	0	0	0	0	0	0	0	0	0	0	0	0
L_B	0	0	0	1	0	0	0	0	0	0	0	0	0	0	0	0	0	0	0	0
C_P	0	0	0	0	1	0	0	0	0	0	0	0	0	0	0	0	0	0	0	0
k	0	0	0	0	0	1	0	0	0	0	0	0	0	0	0	0	0	0	0	0
h	0	0	0	0	0	0	1	0	0	0	0	0	0	0	0	0	0	0	0	0
μ	0	0	0	0	0	0	0	1	0	0	0	0	0	0	0	0	0	0	0	0
μ_W	0	0	0	0	0	0	0	0	1	0	0	0	0	0	0	0	0	0	0	0
ρ	0	0	0	0	0	0	0	0	0	1	0	0	0	0	0	0	0	0	0	0
ΔT	0	0	0	0	0	0	0	0	0	0	1	0	0	0	0	0	0	0	0	0
Δt	0	0	0	0	0	0	0	0	0	0	0	1	0	0	0	0	0	0	0	0
N	0	0	0	0	0	0	0	0	0	0	0	0	1	0	0	0	0	0	0	0
$C_{P,HM}$	0	0	0	0	0	0	0	0	0	0	0	0	0	1	0	0	0	0	0	0
k_{HM}	0	0	0	0	0	0	0	0	0	0	0	0	0	0	1	0	0	0	0	0
μ_{HM}	0	0	0	0	0	0	0	0	0	0	0	0	0	0	0	1	0	0	0	0
ρ_{HM}	0	0	0	0	1	0	0	−1	−1	−1	−1	0	0	1	0	−1	0	1	0	1
D_C	−1	−1	−1	−1	0	−1	0	−1	−1	0	0	−1	1	0	−1	−1	1	3	1	0
v_{HM}	0	0	0	0	1	0	0	−1	−1	0	−3	1	−1	1	0	−1	0	0	−1	3
h_{HM}	0	0	0	0	−1	−1	−1	0	0	0	1	0	0	−1	−1	0	0	0	0	−1

Some of the above dimensionless parameters are recognizable; for example, Π_1, Π_2, Π_3, and Π_4 are geometric ratios for the heated mixer. Π_{15} is the inverse Nusselt number in terms of coil diameter for the heating medium, and Π_{16} is the inverse Reynolds number for the heating medium. The remaining dimensionless parameters contain mixed variables, that is, variables for the stirred fluid and variables for the heating medium. Our first task is to create a set of dimensionless parameters in which each parameter contains only stirred fluid variables or heating medium variables. Table 5.3 presents one such possible basis set for scaling a heated, insulated batch mixer. Note that in Table 5.3, we have divided Π_{15} by Π_3 to obtain an inverse Nusselt number in terms of coil length because heat moves into or from the coil along its length.

We must now determine the heat transfer coefficients for the heating medium and for the stirred fluid. Again, we use dimensional analysis to establish correlations for determining these heat transfer coefficients. First, consider the stirred, that is, agitated, fluid inside the mixer. The geometric variables are vessel diameter D [L] and agitator blade length L_B [L]. The material variables for the heated fluid inside the vessel are heat capacity C_P [$L^2T^{-2}\theta^{-1}$], heat conductivity k [$LMT^{-3}\theta^{-1}$], convective heat transfer coefficient h [$MT^{-3}\theta^{-1}$],

Table 5.3 Basis Set of Dimensionless Parameters for Scaling a Heated, Insulated Batch Mixer

$\Pi_1 = \dfrac{D}{D_C}$	$\Pi_2 = \dfrac{L}{D_C}$	$\dfrac{\Pi_2}{\Pi_3} = \dfrac{L}{L_C}$	$\dfrac{\Pi_4}{\Pi_1} = \dfrac{L_B}{D}$
$\dfrac{\Pi_6}{\Pi_1\Pi_7} = \dfrac{k}{Dh} = Nu^{-1}$	$\Pi_7 = \dfrac{h}{h_{HM}}$	$\dfrac{\Pi_8}{\Pi_9} = \dfrac{\mu}{\mu_W}$	$\Pi_{10} = \dfrac{\rho_{SF}}{\rho_{HM}}$
$\Pi_{11} = \dfrac{h_{HM}\Delta T}{\rho_{HM} v_{HM}^3}$	$\Pi_{12}\Pi_{13} = N\Delta t$	$\dfrac{\Pi_4^2\Pi_{10}\Pi_{13}}{\Pi_8} = \dfrac{\rho L_B^2 N}{\mu}$	$\dfrac{\Pi_{14}\Pi_{16}}{\Pi_{15}} = \dfrac{C_{P,HM}\mu_{HM}}{k_{HM}} = Pr_{HM}$
$\dfrac{\Pi_{14}}{\Pi_{15}} = \dfrac{C_{P,HM}D_C\rho_{HM}}{k_{HM}} = Pe_{HM}$	$\Pi_{15} = \dfrac{k_{HM}}{D_C h_{HM}} = Nu_{HM}^{-1};$	$\Pi_{16} = \dfrac{\mu_{HM}}{\rho_{HM} D_C v_{HM}} = Re_{HM}^{-1};$	$\dfrac{\Pi_1^2\Pi_5\Pi_{10}\Pi_{13}}{\Pi_6} = \dfrac{C_P D^2 \rho N}{k} = Pe$

fluid density ρ [ML^{-3}], bulk fluid viscosity μ [$L^{-1}MT^{-1}$], and wall fluid viscosity μ_W [$L^{-1}MT^{-1}$]. The process variable is agitator rotations n [T^{-1}]. The Dimension Table for the stirred fluid is

Variable		D	L_B	n	μ	μ_W	ρ	C_P	k	h
Dimension	L	1	1	0	−1	−1	−3	2	1	0
	M	0	0	0	1	1	1	0	1	1
	T	0	0	−1	−1	−1	0	−2	−3	−3
	θ	0	0	0	0	0	0	−1	−1	−1

The Dimension matrix for the stirred fluid is

$$\begin{bmatrix} 1 & 1 & 0 & -1 & -1 & -3 & 2 & 1 & 0 \\ 0 & 0 & 0 & 1 & 1 & 1 & 0 & 1 & 1 \\ 0 & 0 & -1 & -1 & -1 & 0 & -2 & -3 & -3 \\ 0 & 0 & 0 & 0 & 0 & 0 & -1 & -1 & -1 \end{bmatrix} \tag{5.32}$$

The Rank matrix is

$$R = \begin{bmatrix} -3 & 2 & 1 & 0 \\ 1 & 0 & 1 & 1 \\ 0 & -2 & -3 & -3 \\ 0 & -1 & -1 & -1 \end{bmatrix} \tag{5.33}$$

which has a determinant of

$$|R| = \begin{vmatrix} -3 & 2 & 1 & 0 \\ 1 & 0 & 1 & 1 \\ 0 & -2 & -3 & -3 \\ 0 & -1 & -1 & -1 \end{vmatrix} = -1 \tag{5.34}$$

The determinant for the Rank matrix is nonsingular; thus the Rank of this Dimension matrix is 4. The number of dimensionless parameters is

$$N_P = N_{Var} - R = 9 - 4 = 5 \tag{5.35}$$

The inverse of R is

$$R^{-1} = \begin{bmatrix} -3 & 2 & 1 & 0 \\ 1 & 0 & 1 & 1 \\ 0 & -2 & -3 & -3 \\ 0 & -1 & -1 & -1 \end{bmatrix}^{-1} = \begin{bmatrix} 0 & 1 & 1 & -2 \\ 0 & 0 & 1 & -3 \\ 1 & 3 & 1 & 0 \\ -1 & -3 & -2 & 2 \end{bmatrix} \quad (5.36)$$

The Bulk matrix is

$$B = \begin{bmatrix} 1 & 1 & 0 & -1 & -1 \\ 0 & 0 & 0 & 1 & 1 \\ 0 & 0 & -1 & -1 & -1 \\ 0 & 0 & 0 & 0 & 0 \end{bmatrix} \quad (5.37)$$

and $-R^{-1} \cdot B$ is

$$-R^{-1} \cdot B = -\begin{bmatrix} 0 & 1 & 1 & -2 \\ 0 & 0 & 1 & -3 \\ 1 & 3 & 1 & 0 \\ -1 & -3 & -2 & 2 \end{bmatrix} \begin{bmatrix} 1 & 1 & 0 & -1 & -1 \\ 0 & 0 & 0 & 1 & 1 \\ 0 & 0 & -1 & -1 & -1 \\ 0 & 0 & 0 & 0 & 0 \end{bmatrix} = \begin{bmatrix} 0 & 0 & 1 & 0 & 0 \\ 0 & 0 & 1 & 1 & 1 \\ -1 & -1 & 1 & -1 & -1 \\ 1 & 1 & -2 & 0 & 0 \end{bmatrix} \quad (5.38)$$

We can now write the Total matrix as

$$T = \begin{array}{c} \\ D \\ L_B \\ N \\ \mu \\ \mu_W \\ \rho \\ C_P \\ k \\ h \end{array} \begin{array}{c} \begin{matrix} \Pi_1 & \Pi_2 & \Pi_3 & \Pi_4 & \Pi_5 & & & & \end{matrix} \\ \begin{bmatrix} 1 & 0 & 0 & 0 & 0 & 0 & 0 & 0 & 0 \\ 0 & 1 & 0 & 0 & 0 & 0 & 0 & 0 & 0 \\ 0 & 0 & 1 & 0 & 0 & 0 & 0 & 0 & 0 \\ 0 & 0 & 0 & 1 & 0 & 0 & 0 & 0 & 0 \\ 0 & 0 & 0 & 0 & 1 & 0 & 0 & 0 & 0 \\ 0 & 0 & 1 & 0 & 0 & 0 & -1 & -1 & 2 \\ 0 & 0 & 1 & 1 & 1 & 0 & 0 & -1 & 3 \\ -1 & -1 & 1 & -1 & -1 & -1 & -3 & -1 & 0 \\ 1 & 1 & -2 & 0 & 0 & 1 & 3 & 2 & -2 \end{bmatrix} \end{array} \quad (5.39)$$

Reading down each of the Π_i columns provides us with the dimensionless parameters for the stirred fluid; they are

$$\Pi_1 = \frac{Dh}{k} = Nu \quad \Pi_2 = \frac{L_B h}{k} \quad \Pi_3 = \frac{\rho n C_P k}{h^2} \quad \Pi_4 = \frac{\mu C_P}{k} = Pr \quad \Pi_5 = \frac{\mu_W C_P}{k}$$

Π_1 is the Nusselt number and Π_4 is the Prandtl number. By dividing Π_2 by Π_1, we get a geometric ratio; namely, $\Pi_2/\Pi_1 = L_B/D$. Dividing Π_4 by Π_5 gives us μ/μ_W, the ratio of the stirred fluid viscosity to the viscosity of the stirred fluid adjacent to the heated wall. The combining Π_2, Π_3, and Π_4 yields a Reynolds-like number for an agitated fluid; namely,

$$\frac{\Pi_2^2 \Pi_3}{\Pi_4} = \frac{\rho L_B^2 n}{\mu} \tag{5.40}$$

Thus the function describing a heated, agitated, fluid in a tank is

$$f\left(Re, Nu, Pr, \frac{\mu}{\mu_W}, \frac{L_B}{D}\right) = 0 \tag{5.41}$$

In terms of the Nusselt number, the solution is

$$Nu = \kappa \cdot f\left(Re, Pr, \frac{\mu}{\mu_W}, \frac{L_B}{D}\right) \tag{5.42}$$

where κ is a constant we determine experimentally.

If we assume a power law relationship between the dependent and independent dimensionless parameters, we can write the solution as

$$Nu = \kappa (Re)^{\alpha} (Pr)^{\beta} \left(\frac{\mu}{\mu_W}\right)^{\chi} \left(\frac{L_B}{D}\right)^{\delta} \tag{5.43}$$

where $Re = \rho L_B^2 N/\mu$. We can determine all the power coefficients in this equation experimentally. If $\chi = \delta = 0$, then we have the equation for heat transfer to a fluid flowing through a pipe. Sieder and Tate proposed $\chi \neq 0$, in which case Eq. (5.43) becomes [6]

$$Nu = \kappa (Re)^{\alpha} (Pr)^{\beta} \left(\frac{\mu}{\mu_W}\right)^{\chi} \tag{5.44}$$

Note that the Sieder–Tate equation is generally presented as [7]

$$Nu = \kappa(Re)^{\alpha}(Pr)^{\beta}\left(\frac{\mu_{W}}{\mu}\right)^{-\chi} \tag{5.45}$$

The difference between Eqs. (5.44) and (5.45) is the sign of χ.

The Sieder–Tate equation has been validated many times since its publication [8,9]. In general, α, β, and χ are stable for a variety of tank or vessel configurations, heating mechanisms, that is, using internal coils or external jacket, with or without baffles, and for a large range of Reynolds numbers. The range of values for α, β, and χ are

$$0.50 \leq \alpha \leq 0.67$$

$$0.25 \leq \beta \leq 0.33$$

$$\chi = -0.14$$

where χ is the power for the dimensionless parameter μ_{W}/μ [8]. κ, however, varies from 0.3 to 1.6, depending upon tank or vessel configuration, heating mechanism, the absence or presence of baffles, Reynolds number, and agitator type. Since each published Sieder–Tate correlation depends upon the above stipulations, it is dangerous to blandly pick one such correlation and use it to scale a tank or vessel. Remember the Sieder–Tate equation depends upon geometric similarity. Scaling a model tank or vessel to a prototype tank or a vessel using a Sieder–Tate equation requires that the model and prototype tank or vessel be geometrically similar.

To determine the heat transfer coefficient for the coil, we use the result of the dimensional analysis done in "Heated, Insulated Batch Mixer." That result was

$$Nu = \kappa \cdot f\left(Pr, Re, \frac{\Pi_1}{\Pi_2}, \frac{\Pi_4^2}{\Pi_1}\right) \tag{5.46}$$

where κ is a constant. The fourth and fifth terms in these functions are generally considered negligible when analyzing such processes. Thus

$$Nu = \kappa \cdot f(Pr, Re) \tag{5.47}$$

We generally assume Eq. (5.47) to be a power law relationship of the form

$$Nu = \kappa \cdot (Re)^{x}(Pr)^{y} \tag{5.48}$$

where x and y are constants determined experimentally. For turbulent flow, x lies between 0.5 and 0.8 and y lies between 0.3 and 0.4; for laminar flow, x is 0.3 and y is 0.3 [3–5].

To scale a heated, insulated batch mixer, we must meet the below criteria

$$
\begin{aligned}
&\Pi_{\mathrm{M}}^{\mathrm{Geometric}} = \Pi_{\mathrm{P}}^{\mathrm{Geometric}} \\
&\Pi_{\mathrm{M}}^{\mathrm{Static}} = \Pi_{\mathrm{P}}^{\mathrm{Static}} \\
&\Pi_{\mathrm{M}}^{\mathrm{Kinematic}} = \Pi_{\mathrm{P}}^{\mathrm{Kinematic}} \\
&\Pi_{\mathrm{M}}^{\mathrm{Dynamic}} = \Pi_{\mathrm{P}}^{\mathrm{Dynamic}} \\
&\Pi_{\mathrm{M}}^{\mathrm{Thermal}} = \Pi_{\mathrm{P}}^{\mathrm{Thermal}}
\end{aligned}
\tag{5.49}
$$

But, before attempting to meet these criteria, we must determine which control regime dominates the model. In this case the possible control regimes are

- laminar flow in the stirred fluid;
- turbulent flow in the stirred fluid;

and

- coil-side convective heat transfer;
- mixer-side convective heat transfer;
- coil wall conductivity.

We must know which of these regimes controls the fluid flow of the mixer and which controls the heat transfer of the mixer before we attempt to scale the mixer.

HEAT EXCHANGERS

Industrially, most heat transfer occurs in heat exchangers. There are a plethora of heat exchanger types, which are classified by flow arrangement and construction type [10]. Shell and tube heat exchangers are the most common heat exchangers in the chemical process industry. Shell and tube heat exchangers consist of a tube bundle placed inside a cylindrical shell. Various types of caps can be placed on the ends of this cylindrical shell. In most cases, the process fluid flows through the tubes of the bundle and the heating or cooling medium flows through the shell, thereby submerging the tube bundle.

Shell and tube heat exchangers tend to be bulky devices which become even more bulky as the need for heat transfer surface area increases. At some point, the need for more heat transfer surface area may require installing additional shell and tube heat exchangers, either in series or in parallel. If this case arises, it may be more economic to use a compacted heat exchanger. Such heat exchangers have high heat transfer surface area to volume ratios. We generally use compact heat exchangers when one of the fluids is a gas, which has a low convective heat transfer coefficient. Compact heat exchangers are classified as flat fin-flat tube, flat fin-circular tube, circular fin-circular tube, and plate-fin, which can be further classified as single pass or multiple pass. In this discussion, we concentrate on shell and tube heat exchangers.

We determine the performance of a given heat exchanger using the equation

$$Q = UA(\Delta T)_{\mathrm{Mean}} \tag{5.50}$$

where Q is the heat transfer rate $[\mathrm{W} = \mathrm{ML^2T^{-3}}]$; U is the overall heat transfer coefficient $[\mathrm{MT^{-3}}\theta^{-1}]$; A is the heat transfer surface area $[\mathrm{L^2}]$; and $(\Delta T)_{\mathrm{Mean}}$ is the mean temperature difference appropriate for the flow configuration of a specified heat exchanger $[\theta]$. We use the mean log temperature difference for shell and tube heat exchangers, which is [11]

$$\Delta T = \frac{\Delta T_2 - \Delta T_1}{\ln(\Delta T_2/\Delta T_1)} \tag{5.51}$$

where, for parallel flow through the shell and tube paths [12]

$$\begin{aligned} \Delta T_1 &= T_{\mathrm{Hot,In}} - T_{\mathrm{Cold,In}} \\ \Delta T_2 &= T_{\mathrm{Hot,Out}} - T_{\mathrm{Cold,Out}} \end{aligned} \tag{5.52}$$

and, for counter-current flow through the shell and tube paths [13]

$$\begin{aligned} \Delta T_1 &= T_{\mathrm{Hot,In}} - T_{\mathrm{Cold,Out}} \\ \Delta T_2 &= T_{\mathrm{Hot,Out}} - T_{\mathrm{Cold,In}} \end{aligned} \tag{5.53}$$

We write the overall heat transfer coefficient as a sum of serial resistances to heat transfer; in other words

$$\frac{1}{U} = \frac{1}{h_{\text{Shell}}} + \frac{1}{h_{\text{Shell,Scale}}} + \frac{1}{(k/x)} + \frac{1}{h_{\text{Tube,Scale}}} + \frac{1}{h_{\text{Tube}}} \tag{5.54}$$

where h_{Shell} and h_{Tube} are the convective heat transfer coefficients for the shell fluid and for the tube fluid, respectively $[\text{MT}^{-3}\theta^{-1}]$; $h_{\text{Shell,Scale}}$ and $h_{\text{Tube,Scale}}$ are the convective heat transfer coefficients for the scale deposited on the shell-side of the equipment and for the scale deposited on the tube-side of the equipment, respectively $[\text{MT}^{-3}\theta^{-1}]$; k is the thermal conductivity of the metal wall through which heat moves $[\text{LMT}^{-3}\theta^{-1}]$; and x is the geometric thickness of the metal wall [L]. Values for $h_{\text{Shell,Scale}}$ and $h_{\text{Tube,Scale}}$ can be found in the open literature [14–17]. If the tubes have fins on them to enhance heat transfer, then each of the convective heat transfer coefficients in Eq. (5.54) will be modified to reflect that additional area. We do not consider finned tubes in this discussion. For new, clean heat exchangers, Eq. (5.54) reduces to

$$\frac{1}{U} = \frac{1}{h_{\text{Shell}}} + \frac{1}{(k/x)} + \frac{1}{h_{\text{Tube}}} \tag{5.55}$$

Let us consider scaling the simplest shell and tube heat exchangers. We assume the model and prototype are geometrically similar. It is a single-pass heat exchanger; in other words, the process fluid in the tubes passes just once through the flooded shell. In such a heat exchanger, the ends of each tube are butt welded to a sieve plate, the number of sieve openings equals the number of tubes in the heat exchanger. The assembled tube bundle is then fitted into a shell and the end plates are welded to the shell, thereby separating the process fluid from the cooling or heating fluid which floods and flows around the tube bundle. A hemispherical cap encloses each end of the shell. Process fluid enters one end cap, flows through tubes of the bundle, then collects in the other end cap, and subsequently discharges into the downstream process. A cooling or heating fluid inlet nozzle is welded to the underside of the shell near the exit end cap, and a cooling or heating fluid discharge nozzle is welded to the top of the shell near the process inlet end cap. Thus the shell flow and tube flow are countercurrent to each other.

The geometric variables are shell diameter D_S [L], shell length L_S [L], shell cross-sectional area A_S $[\text{L}^2]$, number of tubes per unit shell cross-sectional area N_{Tubes} $[\text{L}^{-2}]$, tube diameter D_T [L], tube length L_T [L], and tube heat transfer surface area A_{Tubes} [L^2]. The material variables for the shell-side fluid are heat capacity $C_{P,S}$ [$L^2T^{-2}\theta^{-1}$], heat conductivity k_S [$LMT^{-3}\theta^{-1}$], convective heat transfer coefficient h_S [$MT^{-3}\theta^{-1}$], bulk fluid viscosity μ_S [$L^{-1}MT^{-1}$], wall fluid viscosity $\mu_{W,S}$ [$L^{-1}MT^{-1}$], and fluid density ρ_S [ML^{-3}]. The material variables for the tube-side fluid are heat capacity $C_{P,T}$ [$L^2T^{-2}\theta^{-1}$], heat conductivity k_T [$LMT^{-3}\theta^{-1}$], convective heat transfer coefficient h_T [$MT^{-3}\theta^{-1}$], viscosity μ_T [$L^{-1}MT^{-1}$], wall fluid viscosity $\mu_{W,T}$ [$L^{-1}MT^{-1}$], and fluid density ρ_T [ML^{-3}]. The material variables are determined at the average temperature of the shell-side fluid and the tube-side fluid, respectively. We define the average temperature as

$$T_{Average} = \frac{T_{In} + T_{Out}}{2} \tag{5.56}$$

The process variables are shell-side fluid velocity v_{Shell} [LT^{-1}], tube-side fluid velocity v_{Tube} [LT^{-1}], shell-side pressure drop per unit length $(dP/L)_S$ with dimension [$ML^{-2}T^{-2}$], and tube-side pressure drop per unit length $(dP/L)_T$with dimension [$ML^{-2}T^{-2}$].

The shell-side fluid velocity depends upon whether the tubes are aligned or staggered. Aligned tubes are set in a parade formation, as shown below

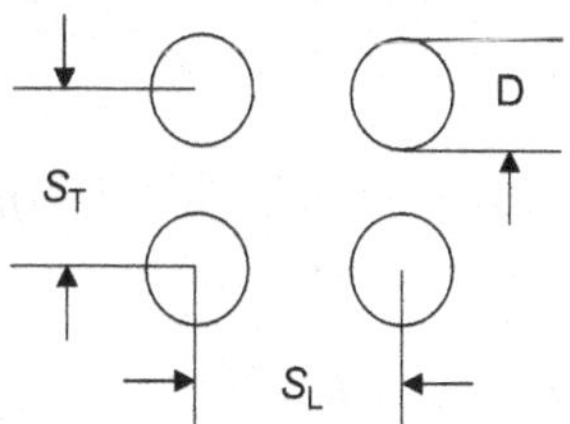

or in a staggered formation, again, as shown below.

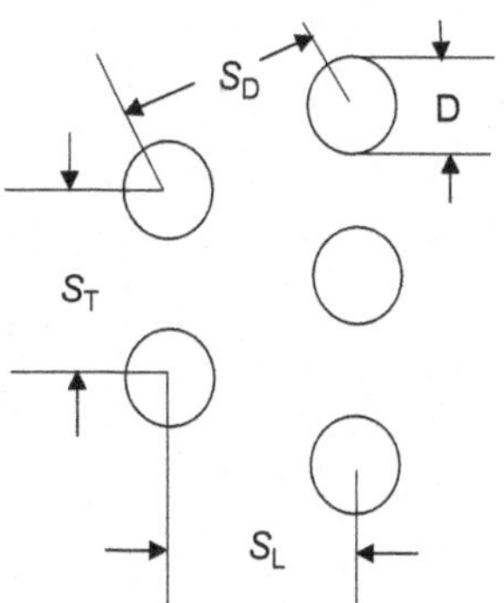

S_T is the transverse pitch, S_L is the longitudinal pitch, and S_D is the diagonal pitch. The shell-side velocity for the aligned formation is

$$v_{\text{Shell,Align}} = \left(\frac{S_T}{S_T - D}\right)v \tag{5.57}$$

where v is the superficial fluid velocity through the shell. The shell-side velocity for the staggered tube formation is

$$v_{\text{Shell,Staggered}} = \left(\frac{S_T}{2(S_D - D)}\right)v \tag{5.58}$$

In this discussion, we assume the simpler tube array, that is, the aligned tube array [18].

Table 5.4 contains the Dimension Table for this scaling example. From the Dimension Table, we can write the Dimension matrix from which we identify the Rank matrix, which is

$$R = \begin{bmatrix} -3 & 2 & 1 & 0 \\ 1 & 0 & 1 & 1 \\ 0 & -2 & -3 & -3 \\ 0 & -1 & -1 & -1 \end{bmatrix} \tag{5.59}$$

The determinant of this Rank matrix is −1; thus its rank is 4. The number of dimensionless parameters is

$$N_P = N_{\text{Var}} - R = 23 - 4 = 19 \tag{5.60}$$

Thus scaling a single-pass heat exchanger requires nineteen dimensionless parameters.

The inverse of R is

$$R^{-1} = \begin{bmatrix} 0 & 1 & 1 & -2 \\ 0 & 0 & 1 & -3 \\ 1 & 3 & 1 & 0 \\ -1 & -3 & -2 & 2 \end{bmatrix} \tag{5.61}$$

Table 5.4 Dimension Table for Scaling a Single-Pass, Tube Aligned, Shell and Tube Heat Exchanger

	D_S	L_S	A_S	N_T	D_T	L_T	A_T	v_S	v_T	$(dP/L)_S$	$(dP/L)_T$	μ_S	$\mu_{S,W}$	$C_{P,S}$	k_S	h_S	ρ_S	μ_T	$\mu_{T,W}$	ρ_T	$C_{P,T}$	k_T	h_T
L	1	1	2	−2	1	1	2	1	1	−2	−2	−1	−1	2	1	0	−3	−1	−1	−3	2	1	0
M	0	0	0	0	0	0	0	0	0	1	1	1	1	0	1	1	1	1	1	1	0	1	1
T	0	0	0	0	0	0	0	−1	−1	−2	−2	−1	−1	−2	−3	−3	0	−1	−1	0	−2	−3	−3
θ	0	0	0	0	0	0	0	0	0	0	0	0	0	−1	−1	−1	0	0	0	0	−1	−1	−1

The Bulk matrix is

$$B=\begin{bmatrix} 1 & 1 & 2 & -2 & 1 & 1 & 2 & 1 & 1 & -2 & -2 & -1 & -1 & 2 & 1 & 0 & -3 & -1 & -1 \\ 0 & 0 & 0 & 0 & 0 & 0 & 0 & 0 & 0 & 1 & 1 & 1 & 1 & 0 & 1 & 1 & 1 & 1 & 1 \\ 0 & 0 & 0 & 0 & 0 & 0 & 0 & -1 & -1 & -2 & -2 & -1 & -1 & -2 & -3 & -3 & 0 & -1 & -1 \\ 0 & 0 & 0 & 0 & 0 & 0 & 0 & 0 & 0 & 0 & 0 & 0 & 0 & -1 & -1 & -1 & 0 & 0 & 0 \end{bmatrix} \tag{5.62}$$

The product of $-R^{-1}\cdot B$ is

$$-R^{-1}\cdot B=\begin{bmatrix} 0 & 0 & 0 & 0 & 0 & 0 & 0 & 1 & 1 & 1 & 1 & 0 & 0 & 0 & 0 & 0 & -1 & 0 & 0 \\ 0 & 0 & 0 & 0 & 0 & 0 & 0 & 1 & 1 & 2 & 2 & 1 & 1 & -1 & 0 & 0 & 0 & 1 & 1 \\ -1 & -1 & -2 & 2 & -1 & -1 & -2 & 0 & 0 & 1 & 1 & -1 & -1 & 0 & -1 & 0 & 0 & -1 & -1 \\ 1 & 1 & 2 & -2 & 1 & 1 & 2 & -1 & -1 & -3 & -3 & 0 & 0 & 0 & 0 & -1 & 0 & 0 & 0 \end{bmatrix} \tag{5.63}$$

We can now write the Total matrix for this scaling effort; we show it in Total Matrix 3. We identify the dimensionless parameters for

Total Matrix 3 The Total Matrix for Scaling a Single-Pass, Tube Aligned, Shell and Tube Heat Exchanger

$T=$	Π_1	Π_2	Π_3	Π_4	Π_5	Π_6	Π_7	Π_8	Π_9	Π_{10}	Π_{11}	Π_{12}	Π_{13}	Π_{14}	Π_{15}	Π_{16}	Π_{17}	Π_{18}	Π_{19}				
D_S	1	0	0	0	0	0	0	0	0	0	0	0	0	0	0	0	0	0	0	0	0	0	0
L_S	0	1	0	0	0	0	0	0	0	0	0	0	0	0	0	0	0	0	0	0	0	0	0
A_S	0	0	1	0	0	0	0	0	0	0	0	0	0	0	0	0	0	0	0	0	0	0	0
N_T	0	0	0	1	0	0	0	0	0	0	0	0	0	0	0	0	0	0	0	0	0	0	0
D_T	0	0	0	0	1	0	0	0	0	0	0	0	0	0	0	0	0	0	0	0	0	0	0
L_T	0	0	0	0	0	1	0	0	0	0	0	0	0	0	0	0	0	0	0	0	0	0	0
A_T	0	0	0	0	0	0	1	0	0	0	0	0	0	0	0	0	0	0	0	0	0	0	0
v_S	0	0	0	0	0	0	0	1	0	0	0	0	0	0	0	0	0	0	0	0	0	0	0
v_T	0	0	0	0	0	0	0	0	1	0	0	0	0	0	0	0	0	0	0	0	0	0	0
$(dP/L)_S$	0	0	0	0	0	0	0	0	0	1	0	0	0	0	0	0	0	0	0	0	0	0	0
$(dP/L)_T$	0	0	0	0	0	0	0	0	0	0	1	0	0	0	0	0	0	0	0	0	0	0	0
μ_S	0	0	0	0	0	0	0	0	0	0	0	1	0	0	0	0	0	0	0	0	0	0	0
$\mu_{S,W}$	0	0	0	0	0	0	0	0	0	0	0	0	1	0	0	0	0	0	0	0	0	0	0
$C_{P,S}$	0	0	0	0	0	0	0	0	0	0	0	0	0	1	0	0	0	0	0	0	0	0	0
k_S	0	0	0	0	0	0	0	0	0	0	0	0	0	0	1	0	0	0	0	0	0	0	0
h_S	0	0	0	0	0	0	0	0	0	0	0	0	0	0	0	1	0	0	0	0	0	0	0
ρ_S	0	0	0	0	0	0	0	0	0	0	0	0	0	0	0	0	1	0	0	0	0	0	0
μ_T	0	0	0	0	0	0	0	0	0	0	0	0	0	0	0	0	0	1	0	0	0	0	0
$\mu_{T,W}$	0	0	0	0	0	0	0	0	0	0	0	0	0	0	0	0	0	0	1	0	0	0	0
ρ_T	0	0	0	0	0	0	0	1	1	1	1	0	0	0	0	0	−1	0	0	0	1	1	−2
$C_{P,T}$	0	0	0	0	0	0	0	1	1	2	1	1	1	−1	0	0	0	1	1	0	0	1	−3
k_T	−1	−1	−2	−2	−1	−1	−2	0	0	1	1	−1	−1	0	−1	0	0	−1	−1	1	3	1	0
h_T	1	1	2	2	1	1	2	−1	−1	−3	−3	0	0	0	0	−1	0	0	0	−1	−3	−2	2

this scaling effort by reading down the first nineteen columns of Total Matrix 3. However, we may choose to use a shortcut method for identifying dimensionless parameters when the Total matrix is large. If we associate the variable of each row of the Total matrix with the appropriate column of the $-R^{-1} \cdot B$ matrix and identify each row of the $-R^{-1} \cdot B$ matrix with its associated variable, then we can read each dimensionless parameter from the $-R^{-1} \cdot B$ matrix. We make these identifications below.

	D_S	L_S	A_S	N_T	D_T	L_T	A_T	v_S	v_T	$(dP/L)_S$	$(dP/L)_T$	μ_S	$\mu_{S,W}$	$C_{P,S}$	k_S	h_S	ρ_S	μ_T	$\mu_{T,W}$
ρ_T	0	0	0	0	0	0	0	1	1	1	1	0	0	0	0	0	−1	0	0
$C_{P,T}$	0	0	0	0	0	0	0	1	1	2	2	1	1	−1	0	0	0	1	1
k_T	−1	−1	−2	−2	−1	−1	−2	0	0	1	1	−1	−1	0	−1	0	0	−1	−1
h_T	1	1	2	2	1	1	2	−1	−1	−3	−3	0	0	0	0	−1	0	0	0

(5.64)

Π_1 is read from the column above labeled D_S; it is

$$\Pi_1 = \frac{D_S h_T}{k_T} \tag{5.65}$$

We obtain Π_2 in a similar manner, by reading down the column labeled L_S; it is

$$\Pi_2 = \frac{L_S h_T}{k_T} \tag{5.66}$$

We derive all the dimensionless parameters in a similar manner. They are

$$\Pi_1 = \frac{D_S h_T}{k_T} \quad \Pi_2 = \frac{L_S h_T}{k_T} \quad \Pi_3 = \frac{A_S h_T^2}{k_T^2} \quad \Pi_4 = \frac{N_T k_T^2}{h_T^2} \quad \Pi_5 = \frac{D_T h_T}{k_T}$$

$$\Pi_6 = \frac{L_T h_T}{k_T} \quad \Pi_7 = \frac{A_T h_T^2}{k_T^2} \quad \Pi_8 = \frac{v_S \rho_T C_{P,T}}{h_T} \quad \Pi_9 = \frac{v_T \rho_T C_{P,T}}{h_T} \quad \Pi_{10} = \frac{(dP/L)_S \rho_T C_{P,T}^2 k_T}{h_T^3}$$

$$\Pi_{11} = \frac{(dP/L)_T \rho_T C_{P,T}^2 k_T}{h_T^3} \quad \Pi_{12} = \frac{\mu_S C_{P,T}}{k_T} \quad \Pi_{13} = \frac{\mu_{S,W} C_{P,T}}{k_T} \quad \Pi_{14} = \frac{C_{P,S}}{C_{P,T}} \quad \Pi_{15} = \frac{k_{P,S}}{k_{P,T}}$$

$$\Pi_{16} = \frac{h_S}{h_T} \quad \Pi_{17} = \frac{\rho_S}{\rho_T} \quad \Pi_{18} = \frac{\mu_T C_{P,T}}{k_T} \quad \Pi_{19} = \frac{\mu_{T,W} C_{P,T}}{k_T}$$

Some of the above dimensionless parameters are recognizable; for example, Π_5 and Π_6 are Nusselt numbers with respect to tube diameter and tube length, respectively. Π_{18} is the Prandtl number for the tube bundle. Π_{19} is the Prandtl number for the fluid at the wall of each tube. The remaining dimensionless parameters contain mixed variables, that is, variables for the shell-side fluid and variables for the tube-side fluid. We must now create a set of dimensionless parameters in which each parameter contains only shell-side fluid variables or tube-side fluid variables. Table 5.5 presents one such possible basis set for scaling a single-pass heat exchanger.

The success of scaling shell and tube heat exchangers depends upon the generalized criteria

$$\Pi_i^{\text{Model}} = \Pi_i^{\text{Prototype}} \tag{5.67}$$

where i represents single or grouped subscripts in Table 5.5. For these criteria to hold, the controlling regime for the prototype must be the same as for the model. When

$$\Pi_i^{\text{Model}} \neq \Pi_i^{\text{Prototype}} \tag{5.68}$$

then we must determine if the controlling regime has changed and if we can restore the equality by using a distorted model, that is

$$(\Pi_i^{\text{Model}})_{\text{Distorted}} = \Pi_i^{\text{Prototype}} \tag{5.69}$$

If the Π_i terms in the inequality $\Pi_i^{\text{Model}} \neq \Pi_i^{\text{Prototype}}$ are important to the success of our scaling effort and if we are unable to develop a viable distortion for obtaining $(\Pi_i^{\text{Model}})_{\text{Distorted}} = \Pi_i^{\text{Prototype}}$, then we must redesign our shell and tube heat exchanger so that $\Pi_i^{\text{Model}} = \Pi_i^{\text{Prototype}}$ for the important dimensionless parameters and document our reasons for dismissing those dimensionless parameters that we consider of lower importance.

After meeting our scaling criteria, as expressed in Eq. (5.67), we must determine h_S and h_T in order to calculate the overall heat transfer coefficient for our prototype. Flow through the shell-side of a shell and tube heat exchanger is quite complex. Initially, fluid enters the shell perpendicular, that is, transverse, to the tube bundle. Upon encountering the tube bundle, some shell-side fluid begins moving longitudinally with respect to the tube bundle. This transition in shell-side flow direction occurs through the tube bundle until the remaining shell-side fluid encounters the shell wall, at which point all the shell-side fluid changes flow direction and moves longitudinally along the tube

Table 5.5 Basis Set of Dimensionless Parameters for Scaling a Single-Pass, Tube Aligned, Shell and Tube Heat Exchanger

$\frac{\Pi_2}{\Pi_1} = \frac{L_S}{D_S}$	$\frac{\Pi_7}{\Pi_3} = \frac{A_T}{A_S}$	$\Pi_3\Pi_4 = N_T A_S$	$\Pi_4\Pi_7 = N_T A_T$	$\frac{\Pi_6}{\Pi_5} = \frac{L_T}{D_T}$
$\Pi_6 = \frac{L_T h_T}{k_T} = Nu_T$	$\frac{\Pi_2\Pi_{16}}{\Pi_{15}} = \frac{L_S h_S}{k_S} = Nu_S$	$\Pi_{18} = \frac{\mu_T C_{P,T}}{k_T} = Pr_T$	$\frac{\Pi_{12}\Pi_{14}}{\Pi_{15}} = \frac{\mu_S C_{P,S}}{k_S} = Pr_S$	$\Pi_6\Pi_9 = \frac{L_T v_T \rho_T C_{P,T}}{k_T}$ $\Pi_6\Pi_9 = Pe_T$
$\frac{\Pi_2\Pi_8\Pi_{14}\Pi_{17}}{\Pi_{15}} = \frac{L_S v_S \rho_S C_{P,S}}{k_S} = Pe_S$	$\frac{\Pi_1\Pi_8\Pi_{17}}{\Pi_{12}} = \frac{\rho_S D_S v_S}{\mu_S} = Re_S$	$\frac{\Pi_5\Pi_9}{\Pi_{18}} = \frac{\rho_T D_T v_T}{\mu_T} = Re_T$	$\frac{\Pi_{11}\Pi_{18}}{\Pi_9^3} = \frac{(dP/L)_T \mu_T}{\rho_T^2 v_T^3}$	$\frac{\Pi_{10}\Pi_{12}}{\Pi_8^3\Pi_{17}^2} = \frac{(dP/L)_S \mu_S}{\rho_S^2 v_S^3}$
$\frac{\Pi_{12}}{\Pi_{13}} = \frac{\mu_S}{\mu_{S,W}}$	$\frac{\Pi_{18}}{\Pi_{19}} = \frac{\mu_T}{\mu_{T,W}}$	$\Pi_6\Pi_9 = \frac{\rho_T L_T v_T C_{P,T}}{k_T} = Gz_T$	$\frac{\Pi_{16}}{\Pi_8\Pi_{14}\Pi_{17}} = \frac{h_S}{\rho_S C_{P,S} v_S} = St_S$	

bundle. Shell-side flow remains parallel to the tube bundle after it changes flow direction, which is not the most efficient way to exchange heat between the shell-side fluid and the tube-side fluid. The most efficient exchange of heat between the two fluids occurs when the shell-side fluid transverses the tube bundle. Shell-side fluid stagnation can also occur in a shell and tube heat exchanger. Little heat transfer occurs from such stagnant fluid. We insert baffles into the shell of a shell and tube heat exchanger to induce transverse flow and to alleviate fluid stagnation.

The types of baffles installed in a shell and tube heat exchanger are

- orifice, which guide shell-side flow parallel to the tube bundle;
- disk and donut, which form annular flow channels that increase the interstitial velocity of the shell-side fluid;
- segmented, which extend alternately from the shell top wall and bottom wall across the tube bundle, thereby inducing shell-side flow transverse to the tube bundle [19].

Staggered baffles ensure the greatest transverse flow of shell-side fluid across the tube bundle. Also, reducing the space between staggered baffles and overlapping them minimizes fluid stagnation within the shell. However, periodically changing shell-side flow direction and forcing the fluid through the tube bundle increases the pressure drop between the entrance and exit nozzles of the shell. Such flow direction changes also make it difficult to mathematically describe the flow pattern inside the shell; and, tubes come in a variety of shapes, thereby increasing the difficulty of describing the flow pattern within the shell [20]. Thus it is best to consider flow through the shell on a macroscopic basis by plotting $(dP/L)_S\mu_S/\rho_S^2 v_S^3$ as a function of shell-side Reynolds number, $\rho_S D_S v_S/\mu_S$ for each heat exchanger being scaled.

We should also consider heat transfer in the shell on a macroscopic basis and determine the average Nusselt number, Nu_S, for shell-side flow. The relationship between Nu_S and the Prandtl number, Pr_S, and the Reynolds number, Re_S, for the shell-side fluid is similar to that determined above for heat transfer in conduits, such as tubes and pipes. A generalized heat transfer correlation for the shell-side fluid is

$$Nu_S = \kappa_S (Re_S)^x (Pr_S)^y \tag{5.70}$$

We do not include the dimensionless term μ/μ_W because of shell-side fluid separation from the tube bundle. The Prandtl number depends only upon material variables; in nearly every process situation, y is 0.33. However, the Reynolds number depends upon shell diameter; thus x must be determined experimentally. The value of κ_S reflects the arrangement of the tubes in the tube bundle, the type of baffles used in the heat exchanger, and the arrangement of the baffles within the heat exchanger shell. Therefore, κ_S must be determined experimentally. Values for κ_S and x are available in the open literature: for single, circular tubes, κ_S ranges from 0.02 to 1.0 and x ranges from 0.33 to 0.81 for different ranges of Re_S [20]. Values for κ_S and x for aligned and staggered tube bundles are also available in the open literature [21–23]. In general, for shell-side flow across tube bundles of nine or more circular tubes, the Nusselt number is [24]

$$Nu_S = 0.32(Re_S)^{0.61}(Pr_S)^{0.31} \tag{5.71}$$

With these relationships, we can calculate the convective heat transfer coefficient h_S for the shell-side fluid, which we then use to calculate the overall heat transfer coefficient for a given shell and tube heat exchanger.

For the tube-side fluid, we determine the convective heat transfer coefficient h_T just as we did earlier for fluid flow in a conduit. For laminar flow in circular tubes, the relationship is

$$Nu_T = \kappa_T(Re_T)^{0.33}(Pr_T)^{0.33}\left(\frac{\mu}{\mu_W}\right)^{0.14} \tag{5.72}$$

and, for turbulent flow, the relationship is

$$Nu_T = \lambda_T(Re_T)^{0.8}(Pr_T)^{y} \tag{5.73}$$

where y is 0.3 for cooling and 0.4 for heating [25]. Note that κ_T and λ_T are constants requiring experimental determination; although, λ_T is generally found to be 0.023. From these relationships, we can calculate the convective heat transfer coefficient h_T for the tube-side fluid, which we then use to calculate the overall heat transfer coefficient for a given shell and tube heat exchanger.

SUMMARY

This chapter presented methods for scaling heat transfer equipment using dimensional analysis. We scaled heated, uninsulated storage tanks; heated, insulated batch mixers; and heat exchangers in this chapter.

REFERENCES

[1] D.Q. Kern, Process Heat Transfer, McGraw-Hill Book Company, New York, NY, 1950, p. 77 and 218.

[2] J. Worstell, Dimensional Analysis, Butterworth-Heinemann, Oxford, UK, 2014, pp. 97–101.

[3] W. McAdams, Heat Transmission, second ed., McGraw-Hill Book Company, New York, NY, 1942, pp. 166–177.

[4] R. Johnstone, M. Thring, Pilot Plants, Models, and Scale-up Methods in Chemical Engineering, McGraw-Hill Book Company, New York, NY, 1957, p. 134.

[5] M. Zlokarnik, Scale-Up in Chemical Engineering, Wiley-VCH Verlag GmbH & Co. KGaA, Weinheim, Germany, 2006, p. 184.

[6] E. Sieder, G. Tate, Heat transfer and pressure drop of liquids in tubes, Ind. Eng. Chem. 28 (1936) 1429.

[7] T. Chilton, T. Drew, R. Jebens, Heat transfer coefficients in agitated vessels, Ind. Eng. Chem. 36 (1944) 510.

[8] F. Holland, F. Chapman, Liquid Mixing and Processing in Stirred Tanks, Reinhold Publishing Corporation, New York, NY, 1966, pp. 152–165.

[9] D. Jo, O. Al-Yahia, R. Altamimi, J. Park, H. Chae, Experimental investigation of convective heat transfer in a narrow rectangular channel for upward and downward flows, Nucl. Eng. Technol. 46 (2014) 195.

[10] S. Hall, Rules of Thumb for Chemical Engineers, fifth ed., Elsevier Inc., Oxford, UK, 2012 (Chapter 2).

[11] D. Kern, Process Heat Transfer, McGraw-Hill Book Company, New York, NY, 1950, pp. 41–43.

[12] F. Incropera, D. DeWitt, Fundamentals of Heat and Mass Transfer, fourth ed., John Wiley & Sons, New York, NY, 1996, p. 590.

[13] F. Incropera, D. DeWitt, Fundamentals of Heat and Mass Transfer, fourth ed., John Wiley & Sons, New York, NY, 1996, p. 591.

[14] F. Incropera, D. DeWitt, Fundamentals of Heat and Mass Transfer, fourth ed., John Wiley & Sons, New York, NY, 1996, p. 585.

[15] Fouling in heat exchange equipment, in: J. Chenoweth, M. Impagliazzo (Eds.), American Society of Mechanical Engineers Symposium, vol. HTD-17, American Society of Mechanical Engineers, New York, NY, 1981.

[16] S. Kakac, A. Bergles, F. Mayinger (Eds.), Heat Exchangers, Hemisphere Publishing, New York, NY, 1981.

[17] S. Kakac, R. Shah, A. Bergles (Eds.), Low Reynolds Number Flow Heat Exchangers, Hemisphere Publishing, New York, NY, 1983.

[18] F. Incropera, D. DeWitt, Fundamentals of Heat and Mass Transfer, fourth ed., John Wiley & Sons, New York, NY, 1996, pp. 378–379.

[19] J. Knudsen, D. Katz, Fluid Dynamics and Heat Transfer, McGraw-Hill Book Company, Inc, New York, NY, 1958, pp. 328–332.

[20] Y. Cengel, Heat Transfer: A Practical Approach, International Edition, McGraw-Hill, New York, NY, 1998, p. 367.

[21] W. McAdams, Heat Transmission, third ed., McGraw-Hill Book Company, Inc, New York, NY, 1954, p. 273.

[22] E. Grimison, Correlation and utilization of new data on flow resistance and heat transfer for cross flow of gases over tube banks, Trans. Am. Soc. Mech. Eng. 59 (1937) 583.

[23] E. Grimison, Correlation and utilization of new data on flow resistance and heat transfer for cross flow of gases over tube banks, Trans. Am. Soc. Mech. Eng. 60 (1938) 273.

[24] A. Lydersen, Fluid Flow and Heat Transfer, John Wiley and Sons, Inc, Chichester, UK, 1979, p. 243.

[25] Y. Cengel, Heat Transfer: A Practical Approach, International Edition, McGraw-Hill, New York, NY, 1998, p. 382.

CHAPTER 6

Scaling Chemical Reactors

INTRODUCTION

Those engineering disciplines concerned with fluid flow, such as aeronautical, civil, and mechanical, have used dimensional analysis to good effect for the past hundred years. Their success is largely attributable to the fact that fluid flow requires only three fundamental dimensions and generates a limited number of dimensionless parameters. Those engineering disciplines can still use the Rayleigh indices method, which is, essentially, a hand calculation, to derive the dimensionless parameters.

Mechanical and chemical engineers are concerned with heat flow, either into or from a given mechanism or process. Working with heat flow requires a fourth fundamental dimension: either temperature or thermal energy, which complicates the algebra of dimensional analysis. The situation is further complicated by the flow of fluid initiated by or required by heat transfer, thereby requiring more dimensionless parameters to fully describe the mechanism or process. And, where complicated algebra occurs, mistakes arise. Dimensional analysis involving four fundamental dimensions has been done many times by hand, but such efforts involve significant amounts of time and effort to obtain the first solution, then to check that solution for possible algebraic errors. Thus the number of applications of dimensional analysis to heat transfer is less than for those situations involving fluid flow only.

The situation is even more complicated for chemical engineers, who are concerned with chemical change and with producing chemical products at acceptable rates. Analyzing chemical processes requires a fifth fundamental dimension, that dimension being "amount of substance," which is moles in the SI system of units. Chemical change also involves fluid flow and heat transfer, either initiated by the chemical reaction itself or required by the chemical process utilizing the reaction to produce a given product. Thus the algebra for dimensional analysis of

Scaling Chemical Processes.
© 2016 Elsevier Inc. All rights reserved.

chemical processes is daunting. Due to the algebraic complexity of the effort, chemical engineers have not utilized dimensional analysis to the extent that other engineering disciplines have utilized it.

The matrix formulation of dimensional analysis and the availability of free-for-use matrix calculators on the Internet solve the algebraic issues for chemical engineers and provide a rapid method for determining the dimensionless parameters best describing a chemical process.

FIRST-ORDER, HOMOGENEOUS BATCH REACTION

When developing a chemical process involving a chemical reaction, our first effort at understanding the chemical reaction is done using laboratory-sized reactors. These reactors are generally round bottom flasks set in heating mantles or submerged in cooling or heating baths. We scale from these laboratory reactors into pilot plant equipment and eventually into commercial plant equipment.

Let us consider a first-order chemical reaction occurring in a laboratory-sized batch reactor submerged in a constant temperature bath. The volume of the reactor is the volume of solvent used during the reaction. The agitator of this batch reactor is so efficient that no concentration differences exist within the process fluid; hence, diffusion of reactant is unimportant. The question is: how do we scale this reaction?

The geometric variables are: reaction volume, that is, the volume of solvent in the reactor V $[L^3]$ and the surface area available for heat transfer A_{Surf} $[L^2]$. The material variables are solvent density ρ $[L^{-3}M]$; solvent viscosity μ $[L^{-1}MT^{-1}]$; solvent heat capacity C_P $[L^2T^{-2}\theta^{-1}]$; solvent heat conductivity k $[LMT^{-3}\theta^{-1}]$; and reactor heat transfer coefficient h $[MT^{-3}\theta^{-1}]$. The process variables are starting concentration of reactant C_S and final concentration of reactant C_F $[L^{-3}N]$, reaction time Δt [T], heat of reaction per unit time and volume $[L^{-1}MT^{-2}]$, where heat of reaction is $[L^2MT^{-2}N^{-1}]$ and C_S is $[L^{-3}N]$—combining these two terms yields the dimension $[L^{-1}MT^{-2}]$. The effective reaction rate constant is k_R $[T^{-1}]$, defined as

$$k_R = k_S e^{-(E/RT_S)} \tag{6.1}$$

where k_S is the reaction rate constant at a specific temperature $[T^{-1}]$; E is the energy of activation for the reaction; and R is the gas constant and the ratio E/R has dimension [θ]. The temperature process variables are ΔT_{WF} [θ], which is the temperature difference between the reaction fluid and the reactor wall, and the starting temperature T_S [θ]. We determine all the physical properties for this reaction at T_S.

The Dimensional Table for a first order, homogeneous batch reactor is

Variable		V	A_{Surf}	Δt	k_S	E/R	ΔT_{WF}	C_F	μ	h	ρ	C_P	k	$C_S\Delta H_R$	T_S	C_S
Dimension	L	3	2	0	0	0	0	−3	−1	0	−3	2	1	−1	0	−3
	M	0	0	0	0	0	0	0	1	1	1	0	1	1	0	0
	T	0	0	1	−1	0	0	0	−1	−3	0	−2	−3	−2	0	0
	θ	0	0	0	0	1	1	0	0	−1	0	−1	−1	0	1	0
	N	0	0	0	0	0	0	1	0	0	0	0	0	0	0	1

and the Dimension matrix is

$$\begin{bmatrix} 3 & 2 & 0 & 0 & 0 & 0 & -3 & -1 & 0 & 2 & 1 & -1 & 0 & -3 \\ 0 & 0 & 0 & 0 & 0 & 0 & 0 & 1 & 1 & 0 & 1 & 1 & 0 & 0 \\ 0 & 0 & 1 & -1 & 0 & 0 & 0 & -1 & -3 & -2 & -3 & -2 & 0 & 0 \\ 0 & 0 & 0 & 0 & 1 & 1 & 0 & 0 & -1 & -1 & -1 & 0 & 1 & 0 \\ 0 & 0 & 0 & 0 & 0 & 0 & 1 & 0 & 0 & 0 & 0 & 0 & 0 & 1 \end{bmatrix} \tag{6.2}$$

The largest square matrix for this Dimension matrix is 5×5; it is

$$R = \begin{bmatrix} 2 & 1 & -1 & 0 & -3 \\ 1 & 1 & 0 & 0 & 0 \\ -2 & -3 & -2 & 0 & 0 \\ -1 & -1 & 0 & 1 & 0 \\ 0 & 0 & 0 & 0 & 1 \end{bmatrix} \tag{6.3}$$

Its determinant is

$$|R| = \begin{vmatrix} 2 & 1 & -1 & 0 & -3 \\ 1 & 1 & 0 & 0 & 0 \\ -2 & -3 & -2 & 0 & 0 \\ -1 & -1 & 0 & 1 & 0 \\ 0 & 0 & 0 & 0 & 1 \end{vmatrix} = -2 \tag{6.4}$$

Thus the Rank of this Dimension matrix is 5. The number of dimensionless parameters is

$$N_{\mathrm{P}} = N_{\mathrm{Var}} - R = 15 - 5 = 10 \tag{6.5}$$

The inverse of R is

$$R^{-1} = \begin{bmatrix} 2 & 1 & -1 & 0 & -3 \\ 1 & 1 & 1 & 0 & 0 \\ -2 & -3 & -2 & 0 & 0 \\ -1 & -1 & 0 & 1 & 0 \\ 0 & 0 & 0 & 0 & 1 \end{bmatrix}^{-1} = \begin{bmatrix} -0.5 & -2.5 & -1 & 0 & -1.5 \\ 1 & 3 & 1 & 0 & 3 \\ -1 & -2 & -1 & 0 & -3 \\ 0.5 & 0.5 & 0 & 1 & 1.5 \\ 0 & 0 & 0 & 0 & 1 \end{bmatrix} \tag{6.6}$$

and the Bulk matrix is

$$B = \begin{bmatrix} 3 & 2 & 0 & 0 & 0 & 0 & -3 & -1 & 0 & -3 \\ 0 & 0 & 0 & 0 & 0 & 0 & 0 & 1 & 1 & 1 \\ 0 & 0 & 1 & -1 & 0 & 0 & 0 & -1 & -3 & 0 \\ 0 & 0 & 0 & 0 & 1 & 1 & 0 & 0 & -1 & 0 \\ 0 & 0 & 0 & 0 & 0 & 0 & 1 & 0 & 0 & 0 \end{bmatrix} \tag{6.7}$$

Therefore, $-R^{-1} \cdot B$ is

$$-R^{-1} \cdot B = \begin{bmatrix} 1.5 & 1 & 1 & -1 & 0 & 0 & 0 & 1 & -0.5 & 1 \\ -3 & -2 & -1 & 1 & 0 & 0 & 0 & -1 & 0 & 0 \\ 3 & 2 & 1 & -1 & 0 & 0 & 0 & 0 & -1 & -1 \\ -1.5 & -1 & 0 & 0 & -1 & -1 & 0 & 0 & 0.5 & 1 \\ 0 & 0 & 0 & 0 & 0 & 0 & -1 & 0 & 0 & 0 \end{bmatrix} \tag{6.8}$$

The Total matrix is presented in Total Matrix 1. The dimensionless parameters are, reading down each Π_i column of the Total matrix

$$\Pi_1 = \frac{VC_{\mathrm{P}}(C_{\mathrm{S}}\Delta H_{\mathrm{R}})^3}{k^3 T_{\mathrm{S}}^{1.5}} \quad \Pi_2 = \frac{A_{\mathrm{Surf}}C_{\mathrm{P}}(C_{\mathrm{S}}\Delta H_{\mathrm{R}})^2}{k^2 T_{\mathrm{S}}} \quad \Pi_3 = \frac{\Delta t C_{\mathrm{P}}(C_{\mathrm{S}}\Delta H_{\mathrm{R}})}{k^3} \quad \Pi_4 = \frac{k_{\mathrm{S}}k}{C_{\mathrm{P}}(C_{\mathrm{S}}\Delta H_{\mathrm{R}})} \quad \Pi_5 = \frac{E/R}{T_{\mathrm{S}}}$$

$$\Pi_6 = \frac{\Delta K_{\mathrm{WF}}}{T_{\mathrm{S}}} \quad \Pi_7 = \frac{C_{\mathrm{F}}}{C_{\mathrm{S}}} \quad \Pi_8 = \frac{\mu C_{\mathrm{P}}}{k} \quad \Pi_9 = \frac{hT_{\mathrm{S}}^{0.5}}{C_{\mathrm{P}}^{0.5}(C_{\mathrm{S}}\Delta H_{\mathrm{R}})} \quad \Pi_{10} = \frac{\rho C_{\mathrm{P}} T_{\mathrm{S}}}{(C_{\mathrm{S}}\Delta H_{\mathrm{R}})}$$

Total Matrix 1 Total Matrix for a Perfectly Mixed, First-Order, Homogeneous Batch Reactor

	Π_1	Π_2	Π_3	Π_4	Π_5	Π_6	Π_7	Π_8	Π_9	Π_{10}					
V	1	0	0	0	0	0	0	0	0	0	0	0	0	0	0
SA	0	1	0	0	0	0	0	0	0	0	0	0	0	0	0
Δt	0	0	1	0	0	0	0	0	0	0	0	0	0	0	0
k_S	0	0	0	1	0	0	0	0	0	0	0	0	0	0	0
E/R	0	0	0	0	1	0	0	0	0	0	0	0	0	0	0
ΔT_{WF}	0	0	0	0	0	1	0	0	0	0	0	0	0	0	0
C_F	0	0	0	0	0	0	1	0	0	0	0	0	0	0	0
μ	0	0	0	0	0	0	0	1	0	0	0	0	0	0	0
h	0	0	0	0	0	0	0	0	1	0	0	0	0	0	0
ρ	0	0	0	0	0	0	0	0	0	1	0	0	0	0	0
C_P	1.5	1	1	−1	0	0	0	1	−0.5	1	−0.5	−2.5	−1	0	−1.5
k	−3	−2	−1	1	0	0	0	−1	0	0	1	3	1	0	3
$C_S\Delta H_R$	3	2	1	−1	0	0	0	0	−1	−1	−1	−2	−1	0	−3
T_S	−1.5	−1	0	0	−1	−1	0	0	0.5	1	0.5	0.5	0	1	1.5
C_S	0	0	0	0	0	0	−1	0	0	0	0	0	0	0	1

The dimensionless analysis of this batch reactor requires ten dimensionless parameters. We immediately recognize that Π_5 is the Arrhenius number; that is, $E/R\ T_S$. Π_7 is the ratio of final reactant concentration to starting reactant concentration—this is the parameter of greatest interest when scaling a batch reactor. Π_8 is the Prandtl number for this unit operation. Π_{10} is an inverted, modified version of the third Damkohler number, which describes the heat liberated as a ratio of bulk heat transfer. Π_6 is the ratio of operating temperatures for the chemical reaction. So, we need six more dimensionless parameters to describe this scaling effort.

The ratio of Π_2^3/Π_1^2 yields A_{Surf}^3/V^2. This dimensionless parameter shows that when we square the reaction volume, we must cube the heat transfer surface area, which explains why batch reactors have a maximum size: there is only so much surface area available for heat transfer. In general, laboratory-sized batch reactors are reaction rate limited because of their large A_{Surf}/V ratios and commercial-sized batch reactors are heat transfer rate limited because of their small A_{Surf}/V ratios.

The product of Π_3 and Π_4 yields a dimensionless time of reaction, namely

$$\Pi_3\Pi_4 = k_S\Delta t \tag{6.9}$$

The ratio

$$\frac{\Pi_2}{\Pi_1\Pi_9} = \frac{A_{\text{Surf}}k}{Vh} \tag{6.10}$$

is a modified, inverse Nusselt number describing this unit operation. The product of Π_2 and Π_4 is

$$\Pi_2\Pi_4 = \frac{A_{\mathrm{Surf}}(C_{\mathrm{S}}\Delta H_{\mathrm{R}})}{kK_{\mathrm{S}}} \tag{6.11}$$

which is a form of the fourth Damkohler number. It describes the ratio of heat liberated to conductive heat transfer.

We need one more dimensionless parameter to form a basis set for this example. We can choose any one of the original set of dimensionless parameters that has whole number indices on the variables. We will choose Π_3 as our tenth dimensionless parameter for this example. Therefore, the solution for this dimensional analysis is

$$f\left(\Pi_3, \Pi_5, \Pi_6, \Pi_7, \Pi_8, \Pi_{10}, \Pi_2\Pi_4, \Pi_3\Pi_4, \frac{\Pi_2^3}{\Pi_1^2}, \frac{\Pi_2}{\Pi_1\Pi_9}\right) = 0 \tag{6.12}$$

In terms of Π_7 or $C_{\mathrm{F}}/C_{\mathrm{S}}$, above function (6.12) can be written as

$$\Pi_7 = \frac{C_{\mathrm{F}}}{C_{\mathrm{S}}} = \kappa \cdot g\left(\Pi_3, \Pi_5, \Pi_6, \Pi_8, \Pi_{10}, \Pi_2\Pi_4, \Pi_3\Pi_4, \frac{\Pi_2^3}{\Pi_1^2}, \frac{\Pi_2}{\Pi_1\Pi_9}\right) \tag{6.13}$$

The criteria for scaling a batch reactor are

$$\begin{aligned}
\Pi_{\mathrm{M}}^{\mathrm{Geometric}} &= \Pi_{\mathrm{P}}^{\mathrm{Geometric}} \\
\Pi_{\mathrm{M}}^{\mathrm{Static}} &= \Pi_{\mathrm{P}}^{\mathrm{Statiic}} \\
\Pi_{\mathrm{M}}^{\mathrm{Kinematic}} &= \Pi_{\mathrm{P}}^{\mathrm{Kinematic}} \\
\Pi_{\mathrm{M}}^{\mathrm{Dynamic}} &= \Pi_{\mathrm{P}}^{\mathrm{Dynamic}} \\
\Pi_{\mathrm{M}}^{\mathrm{Thermal}} &= \Pi_{\mathrm{P}}^{\mathrm{Thermal}} \\
\Pi_{\mathrm{M}}^{\mathrm{Chemical}} &= \Pi_{\mathrm{P}}^{\mathrm{Chemical}}
\end{aligned} \tag{6.14}$$

where the subscript M indicates Model and the subscript P indicates Prototype. The equality $\Pi_{\mathrm{M}}^{\mathrm{Chemical}} = \Pi_{\mathrm{P}}^{\mathrm{Chemical}}$ is represented by the below set of dimensionless parameters.

$\Pi_5 = \frac{E/R}{T_{\mathrm{S}}}$	$\Pi_6 = \frac{\Delta T_{\mathrm{WF}}}{T_{\mathrm{S}}}$	$\Pi_7 = \frac{C_{\mathrm{F}}}{C_{\mathrm{S}}}$	$\Pi_8 = \frac{\mu C_{\mathrm{P}}}{k}$
$\Pi_{10} = \frac{\rho C_{\mathrm{P}} T_{\mathrm{S}}}{(C_{\mathrm{S}}\Delta H_{\mathrm{R}})}$	$\frac{\Pi_2^3}{\Pi_1^2} = \frac{A_{\mathrm{Surf}}^3}{V^2}$	$\Pi_2\Pi_4 = \frac{A_{\mathrm{Surf}}(C_{\mathrm{S}}\Delta H_{\mathrm{R}})k_{\mathrm{S}}}{kT_{\mathrm{S}}}$	$\Pi_3\Pi_4 = k_{\mathrm{S}}\Delta t$
$\frac{\Pi_2}{\Pi_1\Pi_9} = \frac{A_{\mathrm{Surf}}k}{Vh}$		$\Pi_3 = \frac{\Delta t C_{\mathrm{P}}(C_{\mathrm{S}}\Delta H_{\mathrm{R}})}{k^3}$	

For $\Pi_{\mathrm{M}}^{\mathrm{Chemical}} = \Pi_{\mathrm{P}}^{\mathrm{Chemical}}$ to hold true, each of the above Model dimensionless parameters must equal their corresponding Prototype dimensionless parameters. At some point in the design effort, the volume of the prototype reactor will become too large for the available heat transfer surface area. When that happens, $\Pi_{\mathrm{M}}^{\mathrm{Chemical}} \neq \Pi_{\mathrm{P}}^{\mathrm{Chemical}}$ which is caused by

$$\left(\frac{A_{\mathrm{Surf}}^3}{V^2}\right)_{\mathrm{Model}} \neq \left(\frac{A_{\mathrm{Surf}}^3}{V^2}\right)_{\mathrm{Prototype}} \tag{6.15}$$

When we reach this inequality, we must provide additional heat transfer surface area for the reactor. This additional heat transfer surface area will be external to the batch reactor.

We now need to determine the reaction rate constant k_{S} at a specific temperature. Consider the generalized chemical reaction

$$a\mathrm{A} + b\mathrm{B} \rightarrow c\mathrm{C} \tag{6.16}$$

For a spherical, laboratory-sized batch reactor, the component mass balances reduce to

$$\begin{aligned} \frac{d[\mathrm{A}]}{dt} &= -R_{\mathrm{A}} \\ \frac{d[\mathrm{B}]}{dt} &= -R_{\mathrm{B}} \\ \frac{d[\mathrm{C}]}{dt} &= R_{\mathrm{C}} \end{aligned} \tag{6.17}$$

where the negative sign indicates reactant consumption; the brackets [ ] indicate concentration with dimension $[\mathrm{L}^{-3}\mathrm{N}]$; and R_{A}, R_{B}, and R_{C} are the reaction rates $[\mathrm{L}^{-3}\mathrm{NT}^{-1}]$ for species A, B, and C, respectively. Historically, we assume that R is a power law with respect to concentration. Therefore the consumption of reactant A is represented as

$$\frac{d[\mathrm{A}]}{dt} = -R_{\mathrm{A}} = -k_{\mathrm{S}}[\mathrm{A}]^x[\mathrm{B}]^y \tag{6.18}$$

If reactant B greatly exceeds reactant A in the batch reactor, we assume the change in [B] is miniscule compared to the change in reactant A and we take [B] as a constant. Therefore, we convert the above rate equation to

$$\frac{d[\mathrm{A}]}{dt} = -k_{\mathrm{S}}^{\mathrm{Pseudo}}[\mathrm{A}]^x \tag{6.19}$$

where $k_S^{Pseudo} = k_S[B]^y$. We next assume that $x = 1$ so that

$$\frac{d[A]}{dt} = -k_S^{Pseudo}[A] \tag{6.20}$$

We make these assumptions in order to achieve an easily integrable differential equation. Rearranging, then integrating Eq. (6.20) yields

$$\begin{aligned} \int_{[A]_{t=0}}^{[A]} \frac{d[A]}{[A]} &= -k_S^{Pseudo} \int_0^t dt \\ \ln\left(\frac{[A]}{[A]_{t=0}}\right) &= -k_S^{Pseudo} t \end{aligned} \tag{6.21}$$

To prove our assumptions valid, we plot the experimental data from the reaction as $\ln([A]/[A]_{t=0})$ against time. The slope of such a plot is k_S^{Pseudo}. If the plot produces a linear relationship between $\ln([A]/[A]_{t=0})$ and time, we say our assumptions are valid and we have "proven" that the reaction follows a first-order mechanism.

But, does a plot constitute proof of an assertion? We measure the correlation between two data sets with a regression coefficient R. A high positive regression coefficient indicates that the high values of a series of datum entries are associated with the high values of a second series of datum entries. Conversely, a high negative R indicates that the high values of a series of datum entries are associated with the low values of a second series of datum entries [1]. In other words, correlation may occur

- when one variable causes a response, although it is not the causation of the response;
- when two variables depend on a common cause;
- when two variables are interdependent;
- when dependency is fortuitous.

Consider Fig. 6.1, which shows a linear correlation between the national employment index and the national wholesale price index for the United States from 1919 through 1935. R^2 is 0.6474, which indicates that this correlation explains 65% of the relationship between these two variables. This relationship suggests that when wholesale prices are high, employment is high; and conversely, when wholesale prices are low, employment is low. This relationship seems reasonable

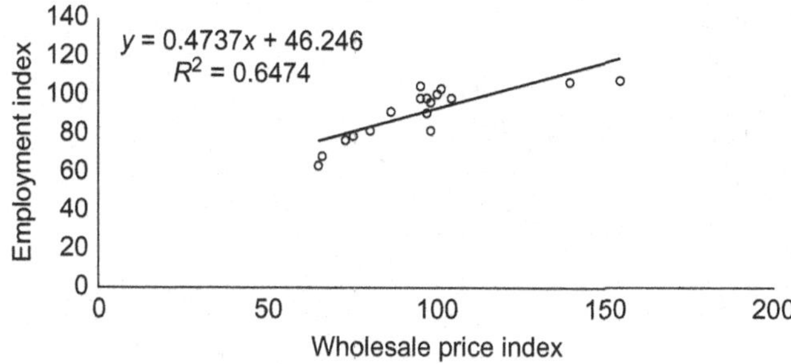

Figure 6.1 Employment index as a function of wholesale price index for 1919 through 1935.

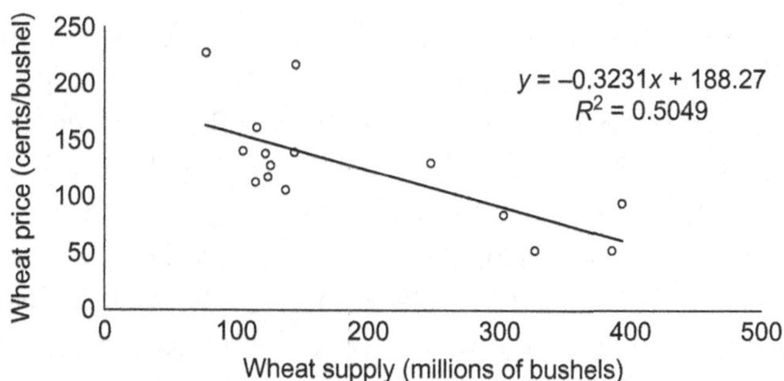

Figure 6.2 Wheat price as a function of wheat supply from 1919 through 1933.

until we realize that both variables depend upon the overall health of the national economy. In other words, when the national economy is "healthy," employment is high and wholesale prices are high; conversely, when the national economy is "unhealthy," employment is low, as are wholesale prices. Thus Fig. 6.1 is an example of two variables depending upon a common cause [2].

Fig. 6.2 shows the relationship between wheat supply in millions of bushels and wheat price in cents/bushel from 1919 through 1933. The correlation is negative; in other words, as wheat production increases, the price of wheat decreases. The correlation is linear and R^2 is 0.5049, indicating that the correlation explains 50% of the variation in the relationship [3]. Fig. 6.2 exemplifies an interdependent relationship, that is, a reciprocal cause and effect relationship. In such relationships, each series, to some extent, is the cause of the other series [4].

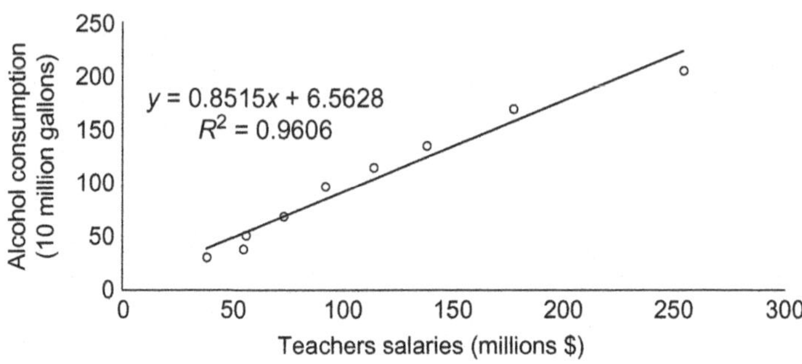

Figure 6.3 Alcohol consumption in the United States as a function of superintendent and teachers salaries from 1870 through 1910.

Consider Fig. 6.3, which shows the consumption of alcohol in the United States, in 10 million gallon increments, as a function of superintendent and teachers' salaries, in million dollar increments, also for the United States. The data is from 1870 through 1910 [5]. A correlation of this information suggests the relationship of these two variables is linear with an R^2 of 0.9606. This correlation explains 96% of the variation contained in these two sets of datum. However, does anyone believe that the alcohol consumed in the United States depends so strongly on superintendent and teachers' salaries? Fig. 6.3 demonstrates that the correlation of two sets of datum can be fortuitous, that is, can be independent of cause and effect.

Our original question was: can a linear correlation of two data sets constitute proof of cause and effect? In other words, can we use a linear correlation as confirmation of an assumption in a chemical kinetic investigation? The conservative answer is "no"; the risqué answer is "maybe."

If a plot does not constitute proof of an assertion, then how do we determine k_S? Consider Fig. 6.4, which is a typical plot for product formation in a batch reactor. Prior to digital computers and user-friendly software, determining the function describing such a plot was labor intensive and time consuming. However, today we can obtain the function describing such plots quite easily. A variety of software packages can be used on our laptop computers to obtain a polynomial function for product formation in a batch reactor, which is

$$[P] = \alpha_P + \beta_P t - \gamma_P t^2 + \delta_P t^3 + \cdots \tag{6.22}$$

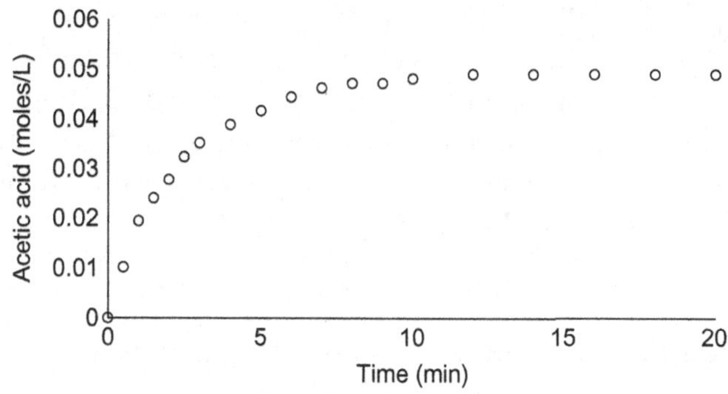

Figure 6.4 Formation of acetic acid by saponification of ethyl acetate.

where [P] is product concentration; t is time; and α_P, β_P, γ_P, and so on are calculated coefficients. The polynomial for reactant concentration is

$$[R] = \alpha_R - \beta_R t + \gamma_R t^2 - \delta_R t^3 + \cdots \tag{6.23}$$

Differentiating the above polynomials gives us

$$\begin{aligned} \frac{d[P]}{dt} &= \beta_P - 2\gamma_P t + 3\delta_P t^2 + \cdots \\ \frac{d[R]}{dt} &= -\beta_R + 2\gamma_R t - 3\delta_R t^2 + \cdots \end{aligned} \tag{6.24}$$

Setting $t = 0$ provides the initial rate of reaction, which is

$$\begin{aligned} \left.\frac{d[P]}{dt}\right|_{t=0} &= \beta_P \\ \left.\frac{d[R]}{dt}\right|_{t=0} &= -\beta_R \end{aligned} \tag{6.25}$$

If we assume a power law relationship for the consumption of reactant A in the above generalized chemical reaction, then

$$\frac{d[A]}{dt} = -k_S[A]^x[B]^y \tag{6.26}$$

and the initial rate of the reaction is

$$\left.\frac{d[A]}{dt}\right|_{t=0} = -\beta = -k_S[A]^x_{t=0}[B]^y_{t=0} \tag{6.27}$$

where $[A]^x_{t=0}$ and $[B]^y_{t=0}$ are the initial concentrations of reactants A and B, respectively. These concentrations are, by the way, the most accurately known concentrations we have when conducting a reaction rate experiment. The concentrations for A and B taken during the reaction are subject to sampling and analytical errors; therefore, taking the natural logarithm of Eq. (6.27) yields

$$\ln \beta = \ln k_S + x \ln[A]_{t=0} + y \ln[B]_{t=0} \tag{6.28}$$

Keeping $[B]_{t=0}$ constant while varying $[A]_{t=0}$ gives us

$$\ln \beta = x \ln[A]_{t=0} + \kappa \tag{6.29}$$

where $\kappa = \ln k_S + y \ln[B]_{t=0}$. Thus a plot of $\ln \beta$ as a function of $[A]_{t=0}$ produces a straight line with slope x, the order of reaction with respect to reactant A. Keeping $[A]_{t=0}$ constant while varying $[B]_{t=0}$ gives us

$$\ln \beta = y \ln[B]_{t=0} + \lambda \tag{6.30}$$

where $\lambda = x \ln[A]_{t=0} + \ln k_S$. Again, plotting $\ln \beta$ as a function of $[B]_{t=0}$ produces a straight line with slope y, the order of reaction with respect to reactant B. Both these plots provide us with a method for calculating k_S. The intercept of each plot is

$$\begin{aligned} i_{\Delta[B]=0} &= \ln k_S + y \ln[B]_{t=0} \\ i_{\Delta[A]=0} &= \ln k_S + y \ln[A]_{t=0} \end{aligned} \tag{6.31}$$

We use these equations to calculate k_S since we know x, y, $\ln[B]_{t=0}$, and $\ln[A]_{t=0}$. This procedure for determining k_S involves only one assumption: the reaction rate is a power law. On the other hand, the historical or traditional method involves a multitude of assumptions in order to calculate k_S.

Consider the saponification of ethyl acetate. The reaction is

$$CH_3COOCH_2CH_3 + NaOH \rightarrow CH_3COONa + CH_3CH_2OH \tag{6.32}$$

Fig. 6.5 displays the decrease of ethyl acetate concentration as a function of time. All these experiments were done at 0.1 moles/L NaOH. We must now decide how to calculate $d[EA]/dt$ using the information shown in Fig. 6.5, where [EA] is ethyl acetate concentration. Historically, we used a straight edge to draw a line tangent to the curve generated by the experiment. We then calculated the slope of that tangent line to obtain, in this case, $d[EA]/dt$. However, we also require

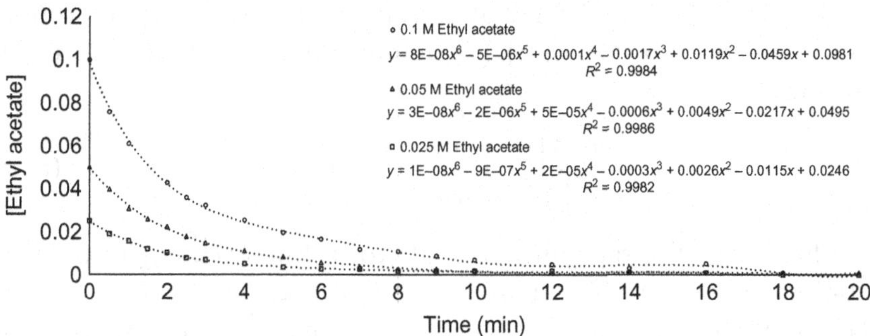

Figure 6.5 [Ethyl acetate] as a function of time at 22°C initial [NaOH] is 0.1 M.

reactant concentration at the time chosen to determine $d[\text{EA}]/dt$. The most accurately known reactant concentration is initial reactant concentration. Therefore, we determined the initial $d[\text{EA}]/dt$ and plotted $\ln\{d[\text{EA}]/dt|_{t=0}\}$ versus $\ln[\text{A}]_{t=0}$. Determining a tangent to a curve by sight involves a significant error, which propagates through the calculations performed during a chemical kinetic analysis. Fortunately, current computing power and user-friendly software obviates such error. Today, we can easily determine the equation describing the relationship between a dependent variable and an independent variable.

Fig. 6.5 displays the equations describing the relationships between ethyl acetate concentration and time. The top equation in Fig. 6.5 is for an initial 0.1 moles/L ethyl acetate; the middle equation is for an initial 0.05 moles/L ethyl acetate; and the bottom equation is for an initial 0.025 moles/L ethyl acetate. From these equations, we can calculate the initial reaction rate for ethyl acetate saponification. For 0.1 moles/L ethyl acetate, the equation is

$$\begin{aligned}[\text{EA}] = 0.0981 - (0.0459)t + (0.0119)t^2 - (0.0017)t^3 \\ + (0.0001)t^4 - (2\times10^{-6})t^5 + (8\times10^{-8})t^6\end{aligned} \tag{6.33}$$

where [EA] is ethyl acetate concentration. Differentiating [EA] with respect to time yields

$$\begin{aligned}\frac{d[\text{EA}]}{dt} = -0.0459 + 2\cdot(0.0119)t - 3\cdot(0.0017)t^2 \\ + 4\cdot(0.0001)t^3 - 5\cdot(2\times10^{-6})t^4 + 6\cdot(8\times10^{-8})t^5\end{aligned} \tag{6.34}$$

Our interest is in the initial rate of ethyl acetate saponification, which occurs at $t = 0$ [6–8]. Therefore, setting $t = 0$ in Eq. (6.34) gives us

$$\left.\frac{d[\mathrm{EA}]}{dt}\right|_{[\mathrm{EA}]=0.1,\ t=0} = -0.0459 \tag{6.35}$$

Thus the initial saponification rate for 0.1 moles/L ethyl acetate is 0.046 moles/L · min. The initial saponification rates for 0.05 and 0.025 moles/L ethyl acetate are easily obtained from the equations shown in Fig. 6.5. Those rates are

$$\left.\frac{d[\mathrm{EA}]}{dt}\right|_{[\mathrm{EA}]=0.05,\ t=0} = -0.0217 \quad \text{and} \quad \left.\frac{d[\mathrm{EA}]}{dt}\right|_{[\mathrm{EA}]=0.025,\ t=0} = -0.0115 \tag{6.36}$$

Table 6.1 tabulates these initial rates and their logarithms, as well as the initial ethyl acetate concentrations and their logarithms. Again, these reaction rates are for an initial [NaOH] of 0.1 moles/L. Fig. 6.6 shows the relationship between $\ln\{d[\mathrm{EA}]/dt|_{t=0}\}$ and $\ln[\mathrm{EA}]_{t=0}$.

Table 6.1 Initial Reactant Concentration and Initial Reaction Rate for Ethyl Acetate Saponification at Constant Sodium Hydroxide Concentration and Variable Ethyl Acetate Concentration

$[\mathrm{EA}]_{t=0}$(moles/L)	$[\mathrm{NaOH}]_{t=0}$(moles/L)	$\left.\frac{d[\mathrm{EA}]}{dt}\right\|_{t=0}$	$\ln[\mathrm{EA}]_{t=0}$	$\ln\left\{\left.\frac{d[\mathrm{EA}]}{dt}\right\|_{t=0}\right\}$
0.1	0.1	0.0459	−2.30	−3.08
0.05	0.1	0.0217	−2.99	−3.83
0.025	0.1	0.0115	−3.68	−4.46

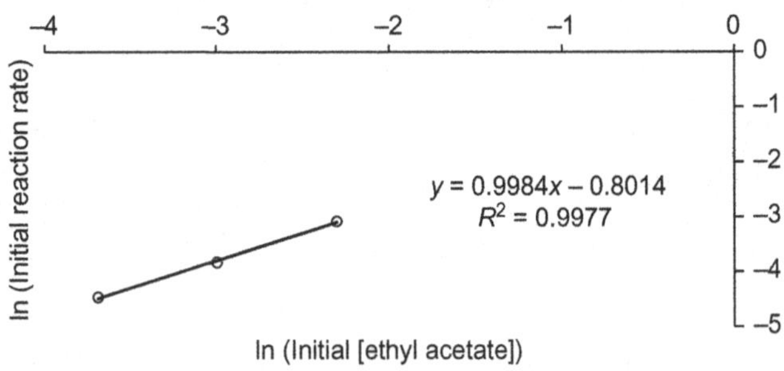

Figure 6.6 ln(Initial reaction rate) as a function of ln (initial [ethyl acetate]). [NaOH] = 0.1 moles/L at 22°C.

The linear correlation of the datum points in Fig. 6.6 shows the relationship of $\ln\{d[EA]/dt|_{t=0}\}$ and $\ln[EA]_{t=0}$ to be linear with an R^2 of 0.9977. The slope of the correlation is 1, which is the order of reaction with respect to ethyl acetate concentration.

Fig. 6.7 shows the relationship between NaOH concentration and time at various NaOH initial concentrations. Ethyl acetate is initially at 0.1 moles/L for each experiment. Fig. 6.7 also displays the equations describing the relationships between NaOH concentration and time at various initial NaOH concentrations. As above, we obtain the initial reaction rates from these equations. Table 6.2 tabulates the initial reaction rates from Fig. 6.7 and their logarithms, as well as the initial NaOH concentrations and their logarithms. Fig. 6.8 shows the relationship between $\ln\{d[NaOH]/dt|_{t=0}\}$ and $\ln[NaOH]_{t=0}$. The linear correlation of the datum points in Fig. 6.8 shows the relationship of $\ln\{d[NaOH]/dt|_{t=0}\}$ and $\ln[NaOH]_{t=0}$ to be linear with an R^2 of 0.9924. The slope of the correlation is 0.94, which is the order of reaction with respect to sodium hydroxide concentration.

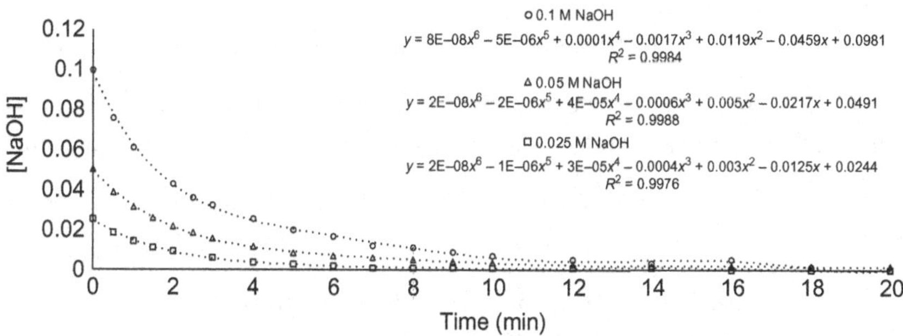

Figure 6.7 [NaOH] as a function of time at 22°C initial [EA] is 0.1 M.

Table 6.2 Initial Reactant Concentration and Initial Reaction Rate for Ethyl Acetate Saponification at Constant Ethyl Acetate Concentration and Variable Sodium Hydroxide Concentration

$[EA]_{t=0}$(moles/L)	$[NaOH]_{t=0}$(moles/L)	$\left.\frac{d[EA]}{dt}\right\|_{t=0}$	$\ln[EA]_{t=0}$	$\ln\left\{\left.\frac{d[EA]}{dt}\right\|_{t=0}\right\}$
0.1	0.1	0.0459	−2.30	−3.08
0.1	0.05	0.0217	−2.99	−3.83
0.1	0.025	0.0125	−3.68	−4.38

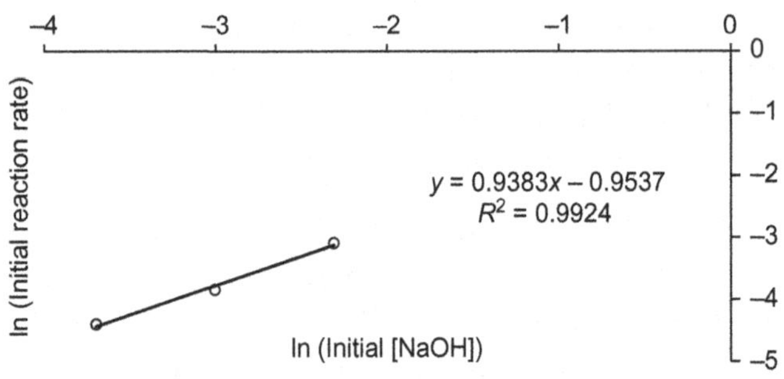

Figure 6.8 ln(Initial reaction rate) as a function of ln (initial [NaOH]). [EA] = 0.1 moles/L at 22° C.

With this information, we can calculate k_{Forward} for ethyl acetate saponification at 22°C. From above, the intercept for the reactions conducted at constant initial NaOH concentration is, per Fig. 6.6

$$\text{Intercept} = y\,\ln[\text{NaOH}]_{t=0} + \ln(k_{\text{Forward}}) \tag{6.37}$$

where y is the slope of the correlation presented in Fig. 6.8. The intercept for the reactions conducted at constant initial ethyl acetate is, from Fig. 6.8

$$\text{Intercept} = x\,\ln[\text{EA}]_{t=0} + \ln(k_{\text{Forward}}) \tag{6.38}$$

where x is the slope of the correlation presented in Fig. 6.6. Therefore the equation for the reactions at 0.1 moles/L NaOH, initially, becomes

$$-0.8014 = (0.9383)^{*}\ln(0.1) + \ln(k_{\text{Forward}}) \tag{6.39}$$

which, upon solving yields $k_{\text{Forward}} = 3.9$. The equation for the reactions at 0.1 moles/L ethyl acetate, initially, becomes

$$-0.9537 = (0.9984)^{*}\ln(0.1) + \ln(k_{\text{Forward}}) \tag{6.40}$$

which, upon solving yields $k_{\text{Forward}} = 3.8$. We can, therefore, write the reaction rate equation for ethyl acetate saponification as

$$-\frac{d[\text{EA}]}{dt} = (3.8)[\text{EA}]^{1}[\text{NaOH}]^{0.94} \tag{6.41}$$

In this procedure, we determined the reaction order of each reactant by changing initial reactant concentration; thus the resulting reaction order is with respect to concentration, which is the true reaction order and we give it the symbol n_c, where $n_c = x_c + y_c$ [9–11]. The concept of reaction order applies only to irreversible reactions or to reactions having a forward reaction rate similar to the equation above. If the reaction is reversible, the reaction order for the forward reaction is generally different from the reaction order for the reverse reaction. Therefore, no overall reaction order exists for reversible reactions since the direction of reaction must be specified when quoting a reaction order [12].

We determine the heat transfer coefficient for this batch reactor using methods discussed in Chapter 5—"Scaling Heat Transfer".

FIRST-ORDER, HOMOGENEOUS BATCH REACTION WITH AGITATION

Let us revisit the above example, but assume imperfect mixing. In other words, we need to include agitation in our analysis of the reaction. Thus we add those variables related to agitation: the power P required to operate the agitator, the shaft rotational speed ω, and the propeller or turbine diameter B. The required power is measured, not calculated. Calculated power contains so many unverifiable assumptions that it is of little use to chemical engineers.

The geometric variables are: reaction volume, that is, the volume of solvent in the reactor V [L^3]; the surface area available for heat transfer A_{Surf} [L^2]; the diameter of the reactor D [L]; and the agitator diameter B [L]. The material variables are solvent density ρ [$L^{-3}M$]; solvent viscosity μ [$L^{-1}MT^{-1}$]; solvent heat capacity C_P [$L^2T^{-2}\theta^{-1}$]; and solvent heat conductivity k [$LMT^{-3}\theta^{-1}$]. The process variables are starting concentration of reactant C_S and final concentration of reactant C_F [$L^{-3}N$], reaction time Δt [T], the power P consumed by the agitator [L^2MT^{-3}], the agitator shaft rotational speed ω [T^{-1}], heat of reaction per unit time and volume [$L^{-1}MT^{-2}$], where heat of reaction is [$L^2MT^{-2}N^{-1}$] and C_S is [$L^{-3}N$]—combining these two terms yields the dimension [$L^{-1}MT^{-2}$]; and reactor heat transfer coefficient h [$MT^{-3}\theta^{-1}$]. The effective reaction rate constant is k_R [T^{-1}], defined as

$$k_R = k_S e^{-(E/RT_S)}$$

where k_S is the reaction rate constant at a specific temperature [T^{-1}], E is the energy of activation for the reaction, and R is the gas constant and the ratio E/R has dimension [θ]. The temperature process variables are ΔT_{WF} [θ], which is the temperature difference between the reaction fluid and the reactor wall, and the starting temperature T_S [θ]. We determine all the physical properties for this reaction at T_S.

Table 6.3 presents the Dimension Table for this scaling effort. The Dimension matrix from that Dimension Table is

$$\begin{bmatrix} 3 & 2 & 1 & 1 & 0 & 0 & 0 & 2 & 0 & 0 & -3 & -1 & 0 & -3 & 2 & 1 & -1 & 0 & -3 \\ 0 & 0 & 0 & 0 & 0 & 0 & 0 & 1 & 0 & 0 & 0 & 1 & 1 & 1 & 0 & 1 & 1 & 0 & 0 \\ 0 & 0 & 0 & 0 & 1 & -1 & -1 & -3 & 0 & 0 & 0 & -1 & -3 & 0 & -2 & -3 & -2 & 0 & 0 \\ 0 & 0 & 0 & 0 & 0 & 0 & 0 & 0 & 1 & 1 & 0 & 0 & -1 & 0 & -1 & -1 & 0 & 1 & 0 \\ 0 & 0 & 0 & 0 & 0 & 0 & 0 & 0 & 0 & 0 & 1 & 0 & 0 & 0 & 0 & 0 & 0 & 0 & 1 \end{bmatrix} \tag{6.42}$$

The largest square matrix for this Dimension matrix is 5 × 5; it is

$$R = \begin{bmatrix} 2 & 1 & -1 & 0 & -3 \\ 1 & 1 & 0 & 0 & 0 \\ -2 & -3 & -2 & 0 & 0 \\ -1 & -1 & 0 & 1 & 0 \\ 0 & 0 & 0 & 0 & 1 \end{bmatrix} \tag{6.43}$$

Its determinant is

$$|R| = \begin{vmatrix} 2 & 1 & -1 & 0 & -3 \\ 1 & 1 & 0 & 0 & 0 \\ -2 & -3 & -2 & 0 & 0 \\ -1 & -1 & 0 & 1 & 0 \\ 0 & 0 & 0 & 0 & 1 \end{vmatrix} = -2 \tag{6.44}$$

Thus the Rank of this Dimension matrix is 5. The number of dimensionless parameters is

$$N_P = N_{Var} - R = 19 - 5 = 14 \tag{6.45}$$

The inverse of R is

$$R^{-1} = \begin{bmatrix} 2 & 1 & -1 & 0 & -3 \\ 1 & 1 & 1 & 0 & 0 \\ -2 & -3 & -2 & 0 & 0 \\ -1 & -1 & 0 & 1 & 0 \\ 0 & 0 & 0 & 0 & 1 \end{bmatrix}^{-1} = \begin{bmatrix} -0.5 & -2.5 & -1 & 0 & -1.5 \\ 1 & 3 & 1 & 0 & 3 \\ -1 & -2 & -1 & 0 & -3 \\ 0.5 & 0.5 & 0 & 1 & 1.5 \\ 0 & 0 & 0 & 0 & 1 \end{bmatrix} \tag{6.46}$$

Table 6.3 Dimension Table for First-Order, Homogenous Batch Reaction With Agitation

Variable		V	A_{Surf}	D	B	Δt	k_S	ω	P	E/R	ΔT_{WF}	C_F	μ	h	ρ	C_P	k	$C_S\Delta H_R$	T_S	C_S
Dimension	L	3	2	1	1	0	0	0	2	0	0	−3	−1	0	−3	2	1	−1	0	−3
	M	0	0	0	0	0	0	0	1	0	0	0	1	1	1	0	1	1	0	0
	T	0	0	0	0	1	−1	−1	−3	0	0	0	−1	−3	0	−2	−3	−2	0	0
	θ	0	0	0	0	0	0	0	0	1	1	0	0	−1	0	−1	−1	0	1	0
	N	0	0	0	0	0	0	0	0	0	0	1	0	0	0	0	0	0	0	1

and the Bulk matrix is

$$B=\begin{bmatrix} 3 & 2 & 1 & 1 & 0 & 0 & 0 & 2 & 0 & 0 & -3 & -1 & 0 & -3 \\ 0 & 0 & 0 & 0 & 0 & 0 & 0 & 1 & 0 & 0 & 0 & 1 & 1 & 1 \\ 0 & 0 & 0 & 0 & 1 & -1 & -1 & -3 & 0 & 0 & 0 & -1 & -3 & 0 \\ 0 & 0 & 0 & 0 & 0 & 0 & 0 & 0 & 1 & 1 & 0 & 0 & -1 & 0 \\ 0 & 0 & 0 & 0 & 0 & 0 & 0 & 0 & 0 & 0 & 1 & 0 & 0 & 0 \end{bmatrix} \tag{6.47}$$

Therefore, $-R^{-1}\cdot B$ is

$$-R^{-1}\cdot B=\begin{bmatrix} 1.5 & 1 & 0.5 & 0.5 & 1 & -1 & -1 & 0.5 & 0 & 0 & 0 & 1 & -0.5 & 1 \\ -3 & -2 & -1 & -1 & -1 & 1 & 1 & -2 & 0 & 0 & 0 & -1 & 0 & 0 \\ 3 & 2 & 1 & 1 & 1 & -1 & -1 & 1 & 0 & 0 & 0 & 0 & -1 & -1 \\ -1.5 & -1 & -0.5 & -0.5 & 0 & 0 & 0 & -1.5 & -1 & -1 & 0 & 0 & 0.5 & 1 \\ 0 & 0 & 0 & 0 & 0 & 0 & 0 & 0 & 0 & 0 & 1 & 0 & 0 & 0 \end{bmatrix} \tag{6.48}$$

We can now assemble the Total matrix, which is shown Total Matrix 2. The dimensionless variables are, reading down the Π_i columns of the Total matrix

$$\Pi_1=\frac{VC_P^{1.5}(C_S\Delta H_R)^3}{k^3T_S^{1.5}}\quad \Pi_2=\frac{A_{Surf}C_P(C_S\Delta H_R)^2}{k^2T_S}\quad \Pi_3=\frac{DC_P^{0.5}(C_S\Delta H_R)}{kT_S^{0.5}}\quad \Pi_4=\frac{BC_P^{0.5}(C_S\Delta H_R)}{kT_S^{0.5}}$$

$$\Pi_5=\frac{\Delta tC_P(C_S\Delta H_R)}{k}\quad \Pi_6=\frac{k_Sk}{C_P(C_S\Delta H_R)}\quad \Pi_7=\frac{\omega k}{C_P(C_S\Delta H_R)}\quad \Pi_8=\frac{PC_P^{0.5}(C_S\Delta H_R)}{k^2T_S^{1.5}}$$

$$\Pi_9=\frac{E/R}{T_S}\quad \Pi_{10}=\frac{\Delta T_{WF}}{T_S}\quad \Pi_{11}=\frac{C_F}{C_S}\quad \Pi_{12}=\frac{\mu C_P}{k}$$

$$\Pi_{13}=\frac{hT_S^{0.5}}{C_P^{0.5}(C_S\Delta H_R)}\quad \Pi_{14}=\frac{\rho C_PT_S}{(C_S\Delta H_R)}$$

Combining dimensionless parameters in order to remove fractional powers gives

$$\Pi_9=\frac{E}{RT_S}=Ar\quad \Pi_{10}=\frac{\Delta T_{WF}}{T_S}\quad \Pi_{11}=\frac{C_F}{C_S}\quad \Pi_{12}=\frac{\mu C_P}{k}=Pr$$

$$\frac{\Pi_1}{\Pi_3^3}=\frac{V}{D^3}\quad \frac{\Pi_2^3}{\Pi_1^2}=\frac{A_{Surf}^3}{V^2}\quad \frac{\Pi_3}{\Pi_4}=\frac{D}{B}\quad \Pi_5\Pi_6=k_S\Delta t$$

$$\Pi_5\Pi_7=\omega\Delta t\quad \Pi_3\Pi_{13}=\frac{Dh}{k}=Nu\quad \frac{\Pi_4^2\Pi_7\Pi_{14}}{\Pi_{12}}=\frac{\rho B^2\omega}{\mu}=Re\quad \Pi_2\Pi_6=\frac{A_{Surf}k_S(C_S\Delta H_R)}{kT_S}=Da^{IV}$$

$$\frac{\Pi_2^3\Pi_5\Pi_6}{\Pi_1^2\Pi_5\Pi_7\Pi_{14}}=\frac{A_{Surf}k_S(C_S\Delta H_R)}{V^2\omega\rho C_PT_S}=Da^{III}\quad \frac{\Pi_8}{\Pi_4}=\frac{P}{BkT_S}$$

Total Matrix 2 The Total Matrix for a First-Order, Homogeneous Batch Reactor With Agitation

	Π_1	Π_2	Π_3	Π_4	Π_5	Π_6	Π_7	Π_8	Π_9	Π_{10}	Π_{11}	Π_{12}	Π_{13}	Π_{14}					
V	1	0	0	0	0	0	0	0	0	0	0	0	0	0	0	0	0	0	0
A_{Surf}	0	1	0	0	0	0	0	0	0	0	0	0	0	0	0	0	0	0	0
D	0	0	1	0	0	0	0	0	0	0	0	0	0	0	0	0	0	0	0
B	0	0	0	1	0	0	0	0	0	0	0	0	0	0	0	0	0	0	0
Δt	0	0	0	0	1	0	0	0	0	0	0	0	0	0	0	0	0	0	0
k_S	0	0	0	0	0	1	0	0	0	0	0	0	0	0	0	0	0	0	0
ω	0	0	0	0	0	0	1	0	0	0	0	0	0	0	0	0	0	0	0
P	0	0	0	0	0	0	0	1	0	0	0	0	0	0	0	0	0	0	0
E/R	0	0	0	0	0	0	0	0	1	0	0	0	0	0	0	0	0	0	0
$T=\Delta T_{WF}$	0	0	0	0	0	0	0	0	0	1	0	0	0	0	0	0	0	0	0
C_F	0	0	0	0	0	0	0	0	0	0	1	0	0	0	0	0	0	0	0
μ	0	0	0	0	0	0	0	0	0	0	0	1	0	0	0	0	0	0	0
h	0	0	0	0	0	0	0	0	0	0	0	0	1	0	0	0	0	0	0
ρ	0	0	0	0	0	0	0	0	0	0	0	0	0	1	0	0	0	0	0
C_P	1.5	1	0.5	0.5	1	−1	−1	0.5	0	0	0	1	−0.5	1	−0.5	−2.5	−1	0	−1.5
k	−3	−2	−1	−1	−1	1	1	−2	0	0	0	−1	0	0	1	3	1	0	3
$C_S\Delta H_R$	3	2	1	1	1	−1	−1	1	0	0	0	0	−1	−1	−1	−2	−1	0	−3
T_S	−1.5	−1	−0.5	−0.5	0	0	0	−1.5	−1	−1	0	0	0.5	1	0.5	0.5	0	1	1.5
C_S	0	0	0	0	0	0	0	0	0	0	−1	0	0	0	0	0	0	0	1

Some of the above dimensionless parameters are readily interpreted. Π_9 is the Arrhenius number and Π_{10} gives the impact of T_S on ΔT_{WF}. Π_{11} indicates the extent of reaction. Π_{12} is the Prandtl number. $\Pi_5\Pi_7$ describes the total rotation during reaction time Δt. $\Pi_5\Pi_6$ is a dimensionless reaction time. $\Pi_3\Pi_{13}$ is the Nusselt number and $\Pi_2\Pi_6$ is the fourth Damkohler number. The third Damkohler number is $\Pi_2^3\Pi_5\Pi_6/\Pi_1^2\Pi_5\Pi_7\Pi_{14}$. $\Pi_4^2\Pi_7\Pi_{14}/\Pi_{12}$ is the Reynolds number in terms of ω. And lastly, Π_8/Π_4 provides a power ratio for the reactor.

The criteria for scaling a batch reactor are

$$\Pi_{M}^{Geometric} = \Pi_{P}^{Geometric}$$

$$\Pi_{M}^{Static} = \Pi_{P}^{Statiic}$$

$$\Pi_{M}^{Kinematic} = \Pi_{P}^{Kinematic}$$

$$\Pi_{M}^{Dynamic} = \Pi_{P}^{Dynamic}$$

$$\Pi_{M}^{Thermal} = \Pi_{P}^{Thermal}$$

$$\Pi_{M}^{Chemical} = \Pi_{P}^{Chemical}$$

where the subscript M indicates Model and the subscript P indicates Prototype. The equality $\Pi_M^{\text{Chemical}} = \Pi_P^{\text{Chemical}}$ is represented by the above set of dimensionless parameters.

For the prototype batch reactor to behave similarly to the model batch reactor, each of the above Model dimensionless parameters must equal its corresponding Prototype dimensionless parameters. Since we have fourteen dimensionless parameters for this scaling effort, we may find that two or more pairs conflict with each other. In other words, we can achieve equality between one pair of model and prototype dimensionless parameters and not achieve equality between a second pair of model and prototype dimensionless parameters. The opposite is also true: if we establish equality between the latter pair of dimensionless parameters, then we cannot achieve equality between the former pair of dimensionless parameters. In such cases, we have to decide which pair is more important for scaling our batch reactor. The other pair of dimensionless parameters is then neglected during the scaling effort. Such decisions must be well documented for future reference and communicated throughout the organization at the time of the decision.

Note that we must determine the specific reaction rate constant k_S as well as the heat transfer coefficient h independent of our scaling effort. We have previously discussed methods for determining k_S and h.

FIRST-ORDER, HOMOGENEOUS REACTION IN A PLUG-FLOW REACTOR [13]

We now consider a homogeneous first-order reaction conduced in a plug-flow, tubular reactor. The process is adiabatic. We will now be concerned with the fluid velocity through the reactor instead of the agitation of the fluid. We will also be concerned with the reactor's diameter D and its length L as well, but not its volume per se. The reaction remains the same: the effective reaction rate constant is k_R $[\mathrm{T}^{-1}]$, defined as

$$k_R = k_S e^{-(E/RT_S)}$$

where E is the energy of activation for the reaction and R is the gas constant and their ratio has dimension $[\theta]$.

The geometric variables are reactor diameter D [L] and reactor length [L]. The material variables are fluid viscosity μ [$L^{-1}MT^{-1}$], fluid density ρ [$L^{-3}M$], fluid heat capacity C_P[$L^2MT^{-2}\theta^{-1}$], and fluid heat conductivity k [$LMT^{-3}\theta^{-1}$]. We must consider the molecular diffusivity of reactant D_{Diff} [L^2T^{-1}] since we assume no back-mixing in the reactor. The process variables are reactant inlet concentration C_{In} and reactant outlet concentration C_{Out} [$L^{-3}N$], heat of reaction $C_{In}\Delta H_R$ [$L^{-1}MT^{-2}$], fluid velocity through the reactor v [LT^{-1}], temperature difference between inlet and outlet fluid ΔT_{IO} [θ], and inlet fluid temperature T_{In} [θ]. We determine all the physical properties for this reaction at T_{In}.

The Dimensional Table is

Variable		L	D	v	D_{Diff}	k_S	E/R	ΔT_{IO}	T_{In}	$C_{In}\Delta H_R$	C_{Out}	C_{In}	ρ	μ	C_P	k
Dimension	L	1	1	1	2	0	0	0	0	−1	−3	−3	−3	−1	2	1
	M	0	0	0	0	0	0	0	0	1	0	0	1	1	1	1
	T	0	0	−1	−1	−1	0	0	0	−2	0	0	0	−1	−2	−3
	θ	0	0	0	0	0	1	1	1	0	0	0	0	0	−1	−1
	N	0	0	0	0	0	0	0	0	0	1	1	0	0	0	0

and the Dimension matrix is

$$\begin{bmatrix} 1 & 1 & 1 & 2 & 0 & 0 & 0 & 0 & -1 & -3 & -3 & -3 & -1 & 2 & 1 \\ 0 & 0 & 0 & 0 & 0 & 0 & 0 & 0 & 1 & 0 & 0 & 1 & 1 & 1 & 1 \\ 0 & 0 & -1 & -1 & -1 & 0 & 0 & 0 & -2 & 0 & 0 & 0 & -1 & -2 & -3 \\ 0 & 0 & 0 & 0 & 0 & 1 & 1 & 1 & 0 & 0 & 0 & 0 & 0 & -1 & -1 \\ 0 & 0 & 0 & 0 & 0 & 0 & 0 & 0 & 0 & 1 & 1 & 0 & 0 & 0 & 0 \end{bmatrix} \tag{6.49}$$

The largest square matrix for this Dimension matrix is 5×5; it is

$$R = \begin{bmatrix} -3 & -3 & -1 & 2 & 1 \\ 0 & 1 & 1 & 1 & 1 \\ 0 & 0 & -1 & -2 & -3 \\ 0 & 0 & 0 & -1 & -1 \\ 1 & 0 & 0 & 0 & 0 \end{bmatrix} \tag{6.50}$$

Its determinant is

$$|R| = \begin{vmatrix} -3 & -3 & -1 & 2 & 1 \\ 0 & 1 & 1 & 1 & 1 \\ 0 & 0 & -1 & -2 & -3 \\ 0 & 0 & 0 & -1 & -1 \\ 1 & 0 & 0 & 0 & 0 \end{vmatrix} = 3 \tag{6.51}$$

Thus the Rank of this Dimension matrix is 5. The number of dimensionless parameters is

$$N_P = N_{Var} - R = 15 - 5 = 10 \tag{6.52}$$

The inverse of R is

$$R^{-1} = \begin{bmatrix} -3 & -3 & -1 & 2 & 1 \\ 0 & 1 & 1 & 1 & 1 \\ 0 & 0 & -1 & -2 & -3 \\ 0 & 0 & 0 & -1 & -1 \\ 1 & 0 & 0 & 0 & 0 \end{bmatrix}^{-1} = \begin{bmatrix} 0 & 0 & 0 & 0 & 1 \\ -0.33 & 0 & 0.33 & -1.33 & -1 \\ 0.33 & 1 & -0.33 & 2.33 & 1 \\ 0.33 & 1 & 0.66 & -0.66 & 1 \\ -0.33 & -1 & -0.66 & -0.33 & -1 \end{bmatrix} \tag{6.53}$$

and the Bulk matrix is

$$B = \begin{bmatrix} 1 & 1 & 1 & 2 & 0 & 0 & 0 & 0 & -1 & -3 \\ 0 & 0 & 0 & 0 & 0 & 0 & 0 & 0 & 0 & 1 \\ 0 & 0 & -1 & -1 & -1 & 0 & 0 & 0 & -2 & 0 \\ 0 & 0 & 0 & 0 & 0 & 1 & 1 & 1 & 0 & 0 \\ 0 & 0 & 0 & 0 & 0 & 0 & 0 & 0 & 0 & 1 \end{bmatrix} \tag{6.54}$$

Therefore, $-R^{-1} \cdot B$ is

$$-R^{-1} \cdot B = \begin{bmatrix} 0 & 0 & 0 & 0 & 0 & 0 & 0 & 0 & 0 & -1 \\ 0.33 & 0.33 & 0.66 & 1 & 0.33 & 1.33 & 1.33 & 1.33 & 0.33 & 0 \\ -0.33 & -0.33 & -0.66 & -1 & -0.33 & -2.33 & -2.33 & -2.33 & -1.33 & 0 \\ -0.33 & -0.33 & 0.33 & 0 & 0.66 & 0.66 & 0.66 & 0.66 & 0.66 & 0 \\ 0.33 & 0.33 & -0.33 & 0 & -0.66 & 0.33 & 0.33 & 0.33 & -0.66 & 0 \end{bmatrix} \tag{6.55}$$

Total Matrix 3 The Total Matrix for a First Oder, Homogeneous Reaction in a Plug-flow Reactor

	Π_1	Π_2	Π_3	Π_4	Π_5	Π_6	Π_7	Π_8	Π_9	Π_{10}					
L	1	0	0	0	0	0	0	0	0	0	0	0	0	0	0
D	0	1	0	0	0	0	0	0	0	0	0	0	0	0	0
v	0	0	1	0	0	0	0	0	0	0	0	0	0	0	0
D_{Diff}	0	0	0	1	0	0	0	0	0	0	0	0	0	0	0
k_S	0	0	0	0	1	0	0	0	0	0	0	0	0	0	0
E/R	0	0	0	0	0	1	0	0	0	0	0	0	0	0	0
ΔK_{IO}	0	0	0	0	0	0	1	0	0	0	0	0	0	0	0
$T =$ K_{In}	0	0	0	0	0	0	0	1	0	0	0	0	0	0	0
$C_{In}\Delta H_R$	0	0	0	0	0	0	0	0	1	0	0	0	0	0	0
C_{Out}	0	0	0	0	0	0	0	0	0	1	0	0	0	0	0
C_{In}	0	0	0	0	0	0	0	0	0	−1					
ρ	0.33	0.33	0.66	1	0.33	1.33	1.33	1.33	0.33	0					
μ	−0.33	−0.33	−0.66	−1	−0.33	−2.33	−2.33	−2.33	−1.33	0			$[R]^{-1}$		
C_P	−0.33	−0.33	0.33	0	0.66	0.66	0.66	0.66	0.66	0					
k	0.33	0.33	−0.33	0	−0.66	0.33	0.33	0.33	−0.66	0					

We can now assemble the Total matrix, as shown in Total Matrix 3. Note that the lower, right portion of the Total matrix is presented as a partitioned matrix; namely, $[R]^{-1}$, in order to save space. That partitioned matrix does not enter the calculation determining the dimensionless variables. The dimensionless variables are, reading down the Π_i columns of Total Matrix 3

$$\Pi_1 = \frac{L\rho^{0.33}k^{0.33}}{\mu^{0.33}C_P^{0.33}} \qquad \Pi_2 = \frac{D\rho^{0.33}k^{0.33}}{\mu^{0.33}C_P^{0.33}} \qquad \Pi_3 = \frac{v\rho^{0.66}C_P^{0.33}}{\mu^{0.66}k^{0.33}}$$

$$\Pi_4 = \frac{D_{Diff}\rho}{\mu} \qquad \Pi_5 = \frac{k_S\rho^{0.33}C_P^{0.66}}{\mu^{0.33}k^{0.66}} \qquad \Pi_6 = \frac{(E/R)\rho^{1.33}C_P^{0.66}}{\mu^{2.33}}$$

$$\Pi_7 = \frac{\Delta T_{IO}\rho^{1.33}C_P^{0.66}k^{0.33}}{\mu^{2.33}} \quad \Pi_8 = \frac{T_{In}\rho^{1.33}C_P^{0.66}k^{0.33}}{\mu^{2.33}} \quad \Pi_9 = \frac{(C_{In}\Delta H_R)\rho^{0.33}C_P^{0.66}}{\mu^{1.33}k^{0.66}}$$

$$\Pi_{10} = \frac{C_{Out}}{C_{In}} \tag{6.56}$$

Combining dimensionless parameters in order to remove fractional powers gives us

$$\frac{\Pi_1}{\Pi_2}=\frac{L}{D} \qquad \Pi_2\Pi_3=\frac{\rho D v}{\mu}=Re \qquad \Pi_4=\frac{\rho D_{\mathrm{Diff}}}{\mu}=Sc^{-1}$$

$$\frac{\Pi_1\Pi_5}{\Pi_3}=\frac{k_S L}{v}=\tau k_S=Da^{\mathrm{I}} \qquad \frac{\Pi_1^2\Pi_5}{\Pi_4}=\frac{L^2 k_S}{D_{\mathrm{Diff}}}=Da^{\mathrm{II}} \qquad \frac{\Pi_1\Pi_5\Pi_9}{\Pi_3\Pi_8}=\frac{(C_{\mathrm{In}}\Delta H_{\mathrm{R}})Lk_S\mu}{\rho v k T_{\mathrm{In}}}$$

$$\frac{\Pi_1^2\Pi_5\Pi_9}{\Pi_8}=\frac{(C_{\mathrm{In}}\Delta H_{\mathrm{R}})L^2 k_S}{k T_{\mathrm{In}}}=Da^{\mathrm{IV}} \qquad \frac{\Pi_6}{\Pi_8}=\frac{E}{RT_{\mathrm{In}}} \qquad \frac{\Pi_7}{\Pi_8}=\frac{\Delta T_{\mathrm{IO}}}{T_{\mathrm{In}}}$$

$$\Pi_{10}=\frac{C_{\mathrm{Out}}}{C_{\mathrm{In}}} \tag{6.57}$$

Most of the above dimensionless parameters are familiar to us. Π_1/Π_2 is the aspect ratio of the tubular reactor. $\Pi_2\Pi_3$ is the Reynolds number. Π_4 is the inverse Schmidt number. $\Pi_1\Pi_5/\Pi_3$ is the Group I Damkohler number, where τ is the average residence time for a reactant molecule in the reactor. $\Pi_1^2\Pi_5/\Pi_4$ is the Group II Damkohler number. Π_6/Π_8 is the Arrhenius dimensionless number and Π_7/Π_8 gives the impact of inlet fluid temperature on ΔT_{IO}. Π_{10} gives the efficiency of the reaction. If we multiply $\Pi_1\Pi_5\Pi_9/\Pi_3\Pi_8$ by $C_{\mathrm{P}}/C_{\mathrm{P}}$; that is, by one, we obtain

$$\frac{\Pi_1\Pi_5\Pi_9}{\Pi_3\Pi_8}\cdot\frac{C_{\mathrm{P}}}{C_{\mathrm{P}}}=\frac{(C_{\mathrm{In}}\Delta H_{\mathrm{R}})Lk_S\mu}{\rho v k T_{\mathrm{In}}}\cdot\frac{C_{\mathrm{P}}}{C_{\mathrm{P}}}=\frac{(C_{\mathrm{In}}\Delta H_{\mathrm{R}})Lk_S}{\rho v C_{\mathrm{P}} T_{\mathrm{In}}}\cdot\frac{\mu C_{\mathrm{P}}}{k}=Da^{\mathrm{III}}\cdot Pr \tag{6.58}$$

where Pr is the Prandtl number, which is the ratio of momentum diffusivity to thermal diffusivity.

The solution for this example is, then

$$f\left(\frac{\Pi_1}{\Pi_2},\Pi_2\Pi_3,\Pi_4,\frac{\Pi_1\Pi_5}{\Pi_3},\frac{\Pi_1^2\Pi_5}{\Pi_4},\frac{\Pi_1\Pi_5\Pi_9}{\Pi_3\Pi_8},\frac{\Pi_1^2\Pi_5\Pi_9}{\Pi_8},\frac{\Pi_6}{\Pi_8},\frac{\Pi_7}{\Pi_8},\Pi_{10}\right)=0 \tag{6.59}$$

Or, in terms of $C_{\mathrm{F}}/C_{\mathrm{S}}$, the solution is

$$\Pi_{10}=\kappa\cdot g\left(\frac{\Pi_1}{\Pi_2},\Pi_2\Pi_3,\Pi_4,\frac{\Pi_1\Pi_5}{\Pi_3},\frac{\Pi_1^2\Pi_5}{\Pi_4},\frac{\Pi_1\Pi_5\Pi_9}{\Pi_3\Pi_8},\frac{\Pi_1^2\Pi_5\Pi_9}{\Pi_8},\frac{\Pi_6}{\Pi_8},\frac{\Pi_7}{\Pi_8}\right) \tag{6.60}$$

The criteria for scaling this particular reactor are the same as for previous examples: the above Model dimensionless parameters must equal their corresponding Prototype dimensionless parameters.

SCALING ADIABATIC FIXED-BED REACTORS

It is difficult to maintain strict similarity in chemical processes due to the number of variables involved. This point is especially true for fixed-bed reactors. For example, we do not change the size or shape of the solid-supported catalyst when scaling a fixed-bed reactor. Thus the ratio of d_P/D, where d_P is the diameter of the catalyst pellet or extrudate and D is the reactor diameter, varies from the model reactor to the prototype reactor. In this case, geometric similarity does not hold when we scale fixed-bed reactors. However, knowledge of this fact should not hinder our use of dimensional analysis when scaling fixed-bed reactors. We simply need to remember it at all times during the scaling effort and make an estimate of its impact on the final effort. d_P/D impacts the distribution of fluid flowing through the fixed-bed reactor. The void fraction of the solid-supported catalyst at the wall is one; therefore, fluid velocity along the wall is high relative to the fluid velocity through the catalyst mass. For large diameter fixed-bed reactors, the fluid velocity profile across the catalyst mass is essentially flat: fluid flow along the reactor wall has little impact on the overall performance of the reactor. However, as D decreases; that is, as d_P/D increases, fluid flow along the reactor wall begins to impact the fluid velocity profile across the catalyst mass, which can impact the overall performance of the fixed-bed reactor [14]. A rule of thumb exists, which states: fixed-bed reactor performance is independent of d_P/D if $D > 10\, d_P$. In other words, fluid flow along the wall of the fixed-bed reactor can be neglected during a scaling effort if the reactor diameter is greater than ten solid-supported catalyst diameters. Such a design will not be strictly similar, geometrically, but it will produce a fixed-bed reactor that performs similarly to its model reactor.

When scaling a fixed-bed reactor, it may be difficult to maintain chemical similarity. The overall reaction rate constant k_O represents

- the movement of reactant molecules through the stagnant film surrounding each solid-supported catalyst pellet or extrudate in the fixed-bed reactor;

- the movement of reactant molecules along a single catalyst pore;
- the conversion of reactant molecules to product molecules at the active site.

If expressed as product formation, k_O represents the above progression in reverse. Note that each of the above steps can be rate-limiting. And, which step is rate-limiting shifts with scale. In general, laboratory-scale fixed-bed reactors are stagnant film diffusion rate limited; pilot plant-scale fixed-bed reactors are stagnant film or pore diffusion rate limited; and commercial-scale fixed-bed reactors are pore diffusion or reaction rate limited. Thus chemical similarity will most likely not hold when scaling a fixed-bed reactor. Again, this fact should not hinder our use of dimensional analysis when scaling a fixed-bed reactor, so long as we remember that the rate controlling step represented by k_O shifts with scale.

Consider a cylindrical tower filled with a solid-supported catalyst. The feed is liquid and flows upward through the reactor. The reactor operates adiabatically. The geometric variables are reactor diameter D [L]; reactor length L [L], which is the height of the catalyst mass; and solid-supported catalyst pellet or extrudate diameter d_P [L]. The material variables are fluid viscosity μ $[L^{-1}MT^{-1}]$, fluid density ρ $[L^{-3}M]$, fluid–solid heat capacity C_P $[L^2MT^{-2}\theta^{-1}]$, fluid–solid heat conductivity k $[LMT^{-3}\theta^{-1}]$, and molecular diffusivity D_{Diff} $[L^2T^{-1}]$. The process variables are

- reactant concentration entering the reactor C_{In} and reactant concentration exiting the reactor C_{Out} $[L^{-3}N]$;
- the heat of reaction $C_{In}\Delta H_R$ $[L^{-1}MT^{-2}]$, where ΔH_R has dimensions of $L^2MT^{-2}N^{-1}$ and C_{In} has dimensions of $[L^{-3}N]$;
- the interstitial fluid velocity through the reactor v $[LT^{-1}]$—interstitial velocity is $v = Q/\varepsilon A$, where Q is volumetric flow rate, A is the cross-sectional area of the empty cylindrical tower, and ε is the void fraction of the porous solid catalyst;
- the fluid temperature entering the reactor T_{In} $[\theta]$—we determine all physical properties using T_{In};
- the temperature difference between the entering fluid and the exiting fluid ΔT_{IO} $[\theta]$;
- the overall rate constant k_O $[T^{-1}]$.

The Dimensional Table is

Variable		L	D	d_P	v	D_{Diff}	k_O	K_{In}	ΔK_{IO}	$C_{In}\Delta H_R$	C_{Out}	C_{In}	ρ	μ	C_P	k
Dimension	L	1	1	1	1	2	0	0	0	−1	−3	−3	−3	−1	2	1
	M	0	0	0	0	0	0	0	0	1	0	0	1	1	1	1
	T	0	0	0	−1	−1	−1	0	0	−2	0	0	0	−1	−2	−3
	θ	0	0	0	0	0	0	1	1	0	0	0	0	0	−1	−1
	N	0	0	0	0	0	0	0	0	0	1	1	0	0	0	0

and the Dimension matrix is

$$\begin{bmatrix} 1 & 1 & 1 & 1 & 2 & 0 & 0 & 0 & -1 & -3 & -3 & -3 & -1 & 2 & 1 \\ 0 & 0 & 0 & 0 & 0 & 0 & 0 & 0 & 1 & 0 & 0 & 1 & 1 & 1 & 1 \\ 0 & 0 & 0 & -1 & -1 & -1 & 0 & 0 & -2 & 0 & 0 & 0 & -1 & -2 & -3 \\ 0 & 0 & 0 & 0 & 0 & 0 & 1 & 1 & 0 & 0 & 0 & 0 & 0 & -1 & -1 \\ 0 & 0 & 0 & 0 & 0 & 0 & 0 & 0 & 0 & 1 & 1 & 0 & 0 & 0 & 0 \end{bmatrix} \tag{6.61}$$

The largest square matrix for this Dimension matrix is 5×5; it is

$$R = \begin{bmatrix} -3 & -3 & -1 & 2 & 1 \\ 0 & 1 & 1 & 1 & 1 \\ 0 & 0 & -1 & -2 & -3 \\ 0 & 0 & 0 & -1 & -1 \\ 1 & 0 & 0 & 0 & 0 \end{bmatrix} \tag{6.62}$$

Its determinant is

$$|R| = \begin{vmatrix} -3 & -3 & -1 & 2 & 1 \\ 0 & 1 & 1 & 1 & 1 \\ 0 & 0 & -1 & -2 & -3 \\ 0 & 0 & 0 & -1 & -1 \\ 1 & 0 & 0 & 0 & 0 \end{vmatrix} = 3 \tag{6.63}$$

Since $|R|$ is 3, the Rank of this Dimension matrix is 5 because it is a 5×5 square matrix. Therefore, number of dimensionless parameters is

$$N_P = N_{Var} - R = 15 - 5 = 10$$

Therefore, this dimensional analysis will produce ten dimensionless parameters.

The inverse of R is

$$R^{-1} = \begin{bmatrix} -3 & -3 & -1 & 2 & 1 \\ 0 & 1 & 1 & 1 & 1 \\ 0 & 0 & -1 & -2 & -3 \\ 0 & 0 & 0 & -1 & -1 \\ 1 & 0 & 0 & 0 & 0 \end{bmatrix}^{-1} = \begin{bmatrix} 0 & 0 & 0 & 0 & 1 \\ -0.33 & 0 & 0.33 & -1.33 & -1 \\ 0.33 & 1 & -0.33 & 2.33 & 1 \\ 0.33 & 1 & 0.66 & -0.66 & 1 \\ -0.33 & -1 & -0.66 & -0.33 & -1 \end{bmatrix} \tag{6.64}$$

and the Bulk Matrix is

$$B = \begin{bmatrix} 1 & 1 & 1 & 1 & 2 & 0 & 0 & 0 & -1 & -3 \\ 0 & 0 & 0 & 0 & 0 & 0 & 0 & 0 & 1 & 0 \\ 0 & 0 & 0 & -1 & -1 & -1 & 0 & 0 & -2 & 0 \\ 0 & 0 & 0 & 0 & 0 & 0 & 1 & 1 & 0 & 0 \\ 0 & 0 & 0 & 0 & 0 & 0 & 0 & 0 & 0 & 1 \end{bmatrix} \tag{6.65}$$

Therefore, $-R^{-1} \cdot B$ is

$$-R^{-1} \cdot B = \begin{bmatrix} 0 & 0 & 0 & 0 & 0 & 0 & 0 & 0 & 0 & -1 \\ 0.33 & 0.33 & 0.33 & 0.66 & 1 & 0.33 & 1.33 & 1.33 & 0.33 & 0 \\ -0.33 & -0.33 & -0.33 & -0.66 & -1 & -0.33 & -2.33 & -2.33 & -1.33 & 0 \\ -0.33 & -0.33 & -0.33 & 0.33 & 0 & 0.66 & 0.66 & 0.66 & 0.66 & 0 \\ 0.33 & 0.33 & 0.33 & -0.33 & 0 & -0.66 & 0.33 & 0.33 & 0.66 & 0 \end{bmatrix} \tag{6.66}$$

Total Matrix 4 contains the dimensionless parameters valid for this reactor. The dimensionless parameters are, reading down the Π_i columns of the Total matrix

$$\Pi_1 = \frac{L\rho^{0.33}k^{0.33}}{\mu^{0.33}C_{\mathrm{P}}^{0.33}} \quad \Pi_2 = \frac{D\rho^{0.33}k^{0.33}}{\mu^{0.33}C_{\mathrm{P}}^{0.33}} \quad \Pi_3 = \frac{d_{\mathrm{P}}\rho^{0.33}k^{0.33}}{\mu^{0.33}C_{\mathrm{P}}^{0.33}}$$

$$\Pi_4 = \frac{v\rho^{0.66}C_{\mathrm{P}}^{0.33}}{\mu^{0.66}k^{0.33}} \quad \Pi_5 = \frac{D_{\mathrm{Diff}}\rho}{\mu} \quad \Pi_6 = \frac{k_{\mathrm{O}}\rho^{0.33}C_P^{0.66}}{\mu^{0.33}k^{0.66}}$$

$$\Pi_7 = \frac{T_{\mathrm{In}}\rho^{1.33}C_{\mathrm{P}}^{0.66}}{\mu^{2.33}} \quad \Pi_8 = \frac{\Delta T_{\mathrm{IO}}\rho^{1.33}C_P^{0.66}}{\mu^{2.33}} \quad \Pi_9 = \frac{(C_{\mathrm{S}}\Delta H_{\mathrm{R}})\rho^{0.33}C_{\mathrm{P}}^{0.66}}{\mu^{1.33}k^{0.66}}$$

$$\Pi_{10} = \frac{C_{\mathrm{Out}}}{C_{\mathrm{In}}} \tag{6.67}$$

Total Matrix 4 The Total Matrix for an Adiabatic Fixed-Bed Reactor

$T =$

	Π_1	Π_2	Π_3	Π_4	Π_5	Π_6	Π_7	Π_8	Π_9	Π_{10}					
L	1	0	0	0	0	0	0	0	0	0	0	0	0	0	0
D	0	1	0	0	0	0	0	0	0	0	0	0	0	0	0
d_P	0	0	1	0	0	0	0	0	0	0	0	0	0	0	0
v	0	0	0	1	0	0	0	0	0	0	0	0	0	0	0
D_{Diff}	0	0	0	0	1	0	0	0	0	0	0	0	0	0	0
k_O	0	0	0	0	0	1	0	0	0	0	0	0	0	0	0
K_{In}	0	0	0	0	0	0	1	0	0	0	0	0	0	0	0
K_{IO}	0	0	0	0	0	0	0	1	0	0	0	0	0	0	0
$C_{In}\Delta H_R$	0	0	0	0	0	0	0	0	1	0	0	0	0	0	0
C_{Out}	0	0	0	0	0	0	0	0	0	1	0	0	0	0	0
C_{In}	0	0	0	0	0	0	0	0	0	−1	0	0	0	0	−1
ρ	0.33	0.33	0.33	0.66	1	0.33	1.33	1.33	0.33	0	0.33	0	−0.33	1.33	1
μ	−0.33	−0.33	−0.33	−0.66	−1	−0.33	−2.33	−2.33	−1.33	0	−0.33	−1	0.33	−2.33	−1
C_P	−0.33	−0.33	−0.33	0.33	0	0.66	0.66	0.66	0.66	0	−0.33	−1	−0.66	0.66	−1
k	0.33	0.33	0.33	−0.33	0	−0.66	0.33	0.33	−0.66	0	0.33	1	0.66	0.33	1

Since the above dimensionless parameters are independent of each other, we can multiply and divide them to remove the fractional indices. Combining dimensionless parameters in order to remove the fractional indices gives

$$\frac{\Pi_1}{\Pi_2} = \frac{L}{D} \qquad \frac{\Pi_3}{\Pi_2} = \frac{d_P}{D} \qquad \Pi_2\Pi_4 = \frac{\rho D v}{\mu} = Re$$

$$\Pi_5 = \frac{\rho D_{Diff}}{\mu} = \mathrm{Sc}^{-1} \qquad \frac{\Pi_1\Pi_6}{\Pi_4} = \frac{Lk_O}{v} = \tau k_O \qquad \frac{\Pi_1^2\Pi_6}{\Pi_5} = \frac{L^2 k_O}{D_{Diff}}$$

$$\frac{\Pi_1\Pi_6\Pi_9}{\Pi_4\Pi_7} = \frac{(C_{In}\Delta H_R)Lk_O}{\rho k v T_{In}} \quad \frac{\Pi_1^2\Pi_6\Pi_9}{\Pi_7} = \frac{(C_{In}\Delta H_R)L^2 k_O}{kT_{In}} \quad \frac{\Pi_8}{\Pi_7} = \frac{\Delta T_{IO}}{T_{In}}$$

$$\Pi_{10} = \frac{C_{Out}}{C_{In}} \tag{6.68}$$

Π_1/Π_2 is the "aspect ratio" of the reactor. Π_3/Π_2 is the ratio of solid-supported catalyst diameter to reactor diameter. $\Pi_2\Pi_4$ is the Reynolds number; Π_5 is the inverse Schmidt number, which is the ratio of momentum diffusivity and molecular diffusivity. $\Pi_1\Pi_6/\Pi_4$ is average residence time that a reactant molecule spends in the reactor, where τ is the average residence time for a reactant molecule in the reactor; it is also the Group I Damkohler number Da^{I}, which is the ratio of chemical reaction rate to bulk flow rate. $\Pi_1^2\Pi_6/\Pi_5$ is the Group II Damkohler number Da^{II}, which is the ratio of chemical reaction rate to molecular to diffusion rate. $\Pi_1^2\Pi_6\Pi_9/\Pi_7$ is the Group IV Damkohler number Da^{IV}, which is the ratio of heat liberated or consumed to conductive heat transfer. If we multiply the dimensionless parameter $\Pi_1\Pi_6\Pi_9/\Pi_4\Pi_7$ by C_P/C_P we obtain

$$\frac{(C_{\text{In}}\Delta H_{\text{R}})k_{\text{O}}L}{\rho v T_{\text{In}} C_{\text{P}}} \cdot \frac{\mu C_{\text{P}}}{k} \tag{6.69}$$

which is the Group III Damkohler number Da^{III} times the Prandtl number Pr. Da^{III} describes the ratio of heat liberated or consumed to the bulk transport of heat and Pr describes momentum diffusivity to thermal diffusivity. Π_8/Π_7 simply tells us how T_{In} impacts ΔT_{IO}. And, Π_{10} provides information about the operating efficiency of the reactor.

The solution for scaling an adiabatic fixed-bed reactor is, by dimensional analysis

$$f\left(\frac{\Pi_1}{\Pi_2}, \frac{\Pi_3}{\Pi_2}, \Pi_2\Pi_4, \Pi_5, \frac{\Pi_1\Pi_6}{\Pi_5}, \frac{\Pi_1^2\Pi_7}{\Pi_5}, \frac{\Pi_1\Pi_6\Pi_9}{\Pi_4\Pi_7}, \frac{\Pi_1^2\Pi_6\Pi_9}{\Pi_7}, \frac{\Pi_8}{\Pi_7}, \Pi_{10}\right) = 0 \tag{6.70}$$

Or, in terms of Π_{10} or $C_{\text{Out}}/C_{\text{In}}$, the solution is

$$\Pi_{10} = \kappa * f\left(\frac{\Pi_1}{\Pi_2}, \frac{\Pi_3}{\Pi_2}, \Pi_2\Pi_4, \Pi_5, \frac{\Pi_1\Pi_6}{\Pi_5}, \frac{\Pi_1^2\Pi_7}{\Pi_5}, \frac{\Pi_1\Pi_6\Pi_9}{\Pi_4\Pi_7}, \frac{\Pi_1^2\Pi_6\Pi_9}{\Pi_7}, \frac{\Pi_8}{\Pi_7}\right) \tag{6.71}$$

For the prototype fixed-bed reactor to operate similarly to the model fixed-bed reactor, the following must hold

$$\left(\frac{C_{\text{Out}}}{C_{\text{In}}}\right)_{\text{Model}} = \left(\frac{C_{\text{Out}}}{C_{\text{In}}}\right)_{\text{Prototype}} \tag{6.72}$$

and all the other dimensionless parameters must equal their counterparts. In other words,

$$\begin{aligned}
&\left(\frac{L}{D}\right)_{\text{Model}} = \left(\frac{L}{D}\right)_{\text{Prototype}} \\
&\left(\frac{d_P}{D}\right)_{\text{Model}} = \left(\frac{d_P}{D}\right)_{\text{Prototype}} \\
&Re_{\text{Model}} = Re_{\text{Prototype}} \\
&Sc^{-1}_{\text{Model}} = Sc^{-1}_{\text{Prototype}} \\
&\left(\frac{Lk_O}{V}\right)_{\text{Model}} = \left(\frac{Lk_O}{V}\right)_{\text{Prototype}} \\
&\left(\frac{L^2k_O}{D_{\text{Diff}}}\right)_{\text{Model}} = \left(\frac{L^2k_O}{D_{\text{Diff}}}\right)_{\text{Prototype}} \\
&\left(\frac{(C_{\text{In}}\Delta H_R)Lk_O}{\rho kvK_{\text{In}}}\right)_{\text{Model}} = \left(\frac{(C_{\text{In}}\Delta H_R)Lk_O}{\rho kvK_{\text{In}}}\right)_{\text{Prototype}} \\
&\left(\frac{(C_{\text{In}}\Delta H_R)L^2k_O}{kK_{\text{In}}}\right)_{\text{Model}} = \left(\frac{(C_{\text{In}}\Delta H_R)L^2k_O}{kK_{\text{In}}}\right)_{\text{Prototype}} \\
&\frac{\Delta K_{IO}}{K_{\text{In}}}\left(\frac{\Delta K_{IO}}{K_{\text{In}}}\right)_{\text{Model}} = \left(\frac{\Delta K_{IO}}{K_{\text{In}}}\right)_{\text{Prototype}}
\end{aligned} \tag{6.73}$$

We know when starting to scale a fixed-bed reactor that

$$\left(\frac{d_P}{D}\right)_{\text{Model}} \neq \left(\frac{d_P}{D}\right)_{\text{Prototype}} \tag{6.74}$$

Thus geometric similarity between the prototype fixed-bed reactor and the model fixed-bed reactor is not strictly held. The question is: can we accept this fact in our prototype design? If we are downscaling a fixed-bed reactor, the fluid flow velocity profile will be flat across the catalyst mass so long as $D > 10\, d_P$. If $D < 10\, d_P$, then a fluid flow velocity profile will develop across the catalyst mass. Thus

$$\left(\frac{C_{Out}}{C_{In}}\right)_{Model} \neq \left(\frac{C_{Out}}{C_{In}}\right)_{Prototype} \tag{6.75}$$

due to a large fraction of the fluid slipping along the reactor wall and not contacting the catalyst mass per se. When upscaling from a small fixed-bed reactor, if $D < 10\, d_P$ for the model, then

$$\left(\frac{C_{Out}}{C_{In}}\right)_{Model} \neq \left(\frac{C_{Out}}{C_{In}}\right)_{Prototype} \tag{6.76}$$

due to a small fraction of the fluid slipping along the scaled reactor wall relative to the fluid slipping along the model reactor wall. In this case, the result, upon operating the prototype fixed-bed reactor will be

$$\left(\frac{C_{Out}}{C_{In}}\right)_{Prototype} \neq \left(\frac{C_{Out}}{C_{In}}\right)_{Model} \tag{6.77}$$

In other words, the scaled fixed-bed reactor will perform better than the reference fixed-bed reactor.

It is unlikely that

$$(Re)_{Model} = (Re)_{Prototype} \tag{6.78}$$

when we scale a fixed-bed reactor. The Reynolds number for laboratory fixed-bed reactors and for pilot plant fixed-bed reactors generally do not equal each other nor do they equal the Reynolds number for a commercial fixed-bed reactor. Therefore, different flow regimes exist at each reactor scale, which means the overall rate constant k_O is different for each reactor scale. Thus

$$\begin{aligned}
\left(\frac{Lk_O}{v}\right)_{Model} &\neq \left(\frac{Lk_O}{v}\right)_{Prototype} \\
\left(\frac{L^2k_O}{D_{Diff}}\right)_{Model} &\neq \left(\frac{L^2k_O}{D_{Diff}}\right)_{Prototype} \\
\left(\frac{(C_{In}\Delta H_R)Lk_O}{\rho kvT_{In}}\right)_{Model} &\neq \left(\frac{(C_{In}\Delta H_R)Lk_O}{\rho kvT_{In}}\right)_{Prototype} \\
\left(\frac{(C_{In}\Delta H_R)L^2k_O}{kT_{In}}\right)_{Model} &\neq \left(\frac{(C_{In}\Delta H_R)L^2k_O}{kT_{In}}\right)_{Prototype}
\end{aligned} \tag{6.79}$$

In this case chemical similarity does not hold between the prototype fixed-bed reactor and the model fixed-bed reactor. But, that does not mean we should abandon dimensional analysis. In fact, the opposite is true: we should use dimensional analysis to determine the extent of the chemical dissimilarity between two fixed-bed reactors. We do that by estimating k_O for the prototype fixed-bed reactor and for the model fixed-bed reactor from Fig. 3.9 in the chapter entitled "Control Regime." We then insert the experimentally determined value into the above dimensionless ratios and compare each pair of dimensionless ratios. If heat required or released is our major concern, then we will want

$$\begin{aligned} \left(\frac{(C_{In}\Delta H_R)Lk_O}{\rho k v T_{In}}\right)_{Model} &\approx \left(\frac{(C_{In}\Delta H_R)Lk_O}{\rho k v T_{In}}\right)_{Prototype} \\ \left(\frac{(C_{In}\Delta H_R)L^2k_O}{kT_{In}}\right)_{Model} &\approx \left(\frac{(C_{In}\Delta H_R)L^2k_O}{kT_{In}}\right)_{Prototype} \end{aligned} \tag{6.80}$$

to be reasonably "close." If these two dimensionless parameters are not "close," then we will have to make a decision about continuing the project or we will have to make a design change. If the reaction under investigation is thermally neutral, then we will not be concerned with consumed or released heat; rather, we will be concerned with reaction performance. In other words, we want

$$\begin{aligned} \left(\frac{Lk_O}{v}\right)_{Model} &\approx \left(\frac{Lk_O}{v}\right)_{Prototype} \\ \left(\frac{L^2k_O}{D_{Diff}}\right)_{Model} &\approx \left(\frac{L^2k_O}{D_{Diff}}\right)_{Prototype} \end{aligned} \tag{6.81}$$

to be "close," if not equal. If these dimensionless parameters are not "close," then we will have to make a decision about continuing the project or we will have to make a design change.

It is difficult when scaling fixed-bed reactors to achieve geometric similarity because

$$\left(\frac{d_P}{D}\right)_{Model} \neq \left(\frac{d_P}{D}\right)_{Prototype} \tag{6.82}$$

and it is difficult to maintain chemical similarity because

$$(Re)_{\text{Model}} = (Re)_{\text{Prototype}}$$

establishes different rate controlling regimes in different sized fixed-bed reactors. By using dimensional analysis, however, we can determine the extent to which a prototype fixed-bed reactor and its model fixed-bed reactor are different. This information provides insight as to why

$$\left(\frac{C_{\text{Out}}}{C_{\text{In}}}\right)_{\text{Model}} \neq \left(\frac{C_{\text{Out}}}{C_{\text{In}}}\right)_{\text{Prototype}}$$

occurs or forecasts that outcome. Knowledge of these differences saves much time, money, and finger-pointing during commissioning and the early operation of a scaled fixed-bed reactor. Time and money are saved because we are not trying to determine why

$$\left(\frac{C_{\text{Out}}}{C_{\text{In}}}\right)_{\text{Model}} \neq \left(\frac{C_{\text{Out}}}{C_{\text{In}}}\right)_{\text{Prototype}}$$

We expected they would not be equal. And, since our expectation was met, the outcome becomes a non-event.

SUMMARY

In this chapter, we applied dimensional analysis to chemical processes. We used homogeneous batch reactions, plug flow reactions, and porous solid–catalyzed reactions as examples. We demonstrated how to derive the dimensionless parameters for these examples, then we showed how to combine them to form the Group I, II, III, and IV Damkohler numbers, as well as the Reynolds number. We demonstrated how these dimensionless parameters and numbers are used during upscaling and downscaling.

REFERENCES

[1] F. Croxton, Elementary Statistics with Applications in Medicine and the Biological Sciences, Dover Publications, Inc, New York, NY, 1959, pp. 120–126.

[2] Anon., Handbook of Labor Statistics, U.S. Government Printing Office, Washington, DC, 1936. p. 132 and 673.

[3] Anon., Yearbook of Agriculture, U.S. Government Printing Office, Washington, DC, 1935, pp. 363–364.

[4] F. Croxton, Elementary Statistics with Applications in Medicine and the Biological Sciences, Dover Publications, Inc, New York, NY, 1959, pp. 120–126.

[5] Anon., Statistical Abstract of the United States, U.S. Government Printing Office, Washington, DC, 1918, pp. 830–835.

[6] Bor-Yea Hsu, University of Houston, personal communication, August 2014.

[7] K. Hall, T. Quickenden, D. Watts, Rate constants from initial concentration data, J. Chem. Educ. 53 (1976) 493.

[8] J. Espensen, Chemical Kinetics and Reaction Mechanisms, second ed., McGraw-Hill Inc, New York, NY, 1995. p. 8.

[9] M. Letort, Definition and determination of the two orders of a chemical reaction, J. Chim. Phys. 34 (1937) 206.

[10] M. Letort, Contribution al'etude du Mecanisme de la reaction chimique, Bull. Soc. Chim. France 9 (1942) 1.

[11] K. Laidler, Chemical Kinetics, second ed., McGraw-Hill Book Company, New York, NY, 1965, pp. 15–17.

[12] K. Laidler, Reaction Kinetics: Volume 1, Homogeneous Gas Reactions, Pergamon Press, Oxford, UK, 1963, pp. 16–19.

[13] M. Zlokarnik, Scale-up in Chemical Engineering, second ed., Wiley-VCH Verlag GmbH & Co., KGaAA, Weinheim, Germany, 2006, pp. 212–214.

[14] M. Tarhan, Catalytic Reactor Design, McGraw-Hill Book Company, 1983, pp. 80–81.

Printed in the United States
By Bookmasters

Printed in the United States
By Bookmasters